"The best remedy for a case of the mid-winter blahs or cabin fever is a good dose of Casselmania."
Joanna Manning, The Tribune

"I read and loved your folk sayings. As an English teacher, I'm always on the lookout for things to read to my class or simply enjoy myself."
Peggy Warren, Spiritwood, Saskatchewan

"I had the pleasure of receiving both your word books as a Christmas present and am still reading and re-reading them."
Guy Charbonneau, Timmins, Ontario

"I enjoyed your Canadian word stories in the books and on CBC Radio."
T.C. Farrell, Head of Chezzetcook, Nova Scotia

"I received your book *Casselmania* and had a thoroughly good read of it over the holidays. Continue to have fun and entertain us all!"
Reino Kokkila, Etobicoke, Ontario

"For a full appreciation of how and why Canadians came by their unique linguistic heritage, there can be no better guide than Bill Casselman."
Moira Farr, Equinox *magazine*

Also available from Little, Brown (Canada)

Casselmania: More Wacky Canadian Words & Sayings

CANADIAN GARDEN WORDS

The Origin of Flower, Tree,
and Plant Names,
both wild and domestic,
entertainingly derived
from their sources
in the Ancient Tongues
together with
Fancy Botanical Names
& Why You Shall Never
Again Be Afraid To Use Them !

BILL CASSELMAN

LITTLE, BROWN AND COMPANY (CANADA) LTD.

BOSTON · NEW YORK · TORONTO · LONDON

CANADIAN CATALOGUING IN PUBLISHING DATA

Casselman, Bill, 1942–
            Canadian garden words

ISBN 0-316-13343-4

1. Botany - Nomenclature. 2. Plant names, Popular - Dictionaries. I. Title.

QK11.C37 1997   580'.1'4   C97-931652-9

Cover and text design by Tania Craan
Cover photograph by First Light/Kunst & Scheidulin
Printed and bound in Canada by Best Book Manufacturers

Little, Brown and Company (Canada) Limited
148 Yorkville Avenue, Toronto, Ontario, M5R 1C2

for Janet Slater
gardener, merry grig, and my friend

CONTENTS

PERENNIALS

TREES NATIVE TO CANADA

WILD PLANTS OF CANADA

WHAT THIS BOOK IS ALL ABOUT

I am a word-nut and an amateur gardener. While blizzards bluster outside my window, the only thing I can plough through are seed catalogues, like many a Canuck greenthumb. Dreaming of spring's surge, of peated loam awaiting seed, of rock-garden nooks for new alpines, I also dawdle over garden volumes that explain how to grow plants. But I could never find a book that offered the juicy lore associated with plant names. Because no particular compilation filled Bill's bill, I made my own. Of course, I use the word *garden* freely. After all, this book includes notes about the names of some Canadian wild flowers and many native trees. But the verdant fields of our dominion, indeed of our earth, are one vast garden. It behooves us to protect it. A step toward that goal is to know better the charm and surprise of its nomenclature.

Browsing at a garden centre, most of us have squinted at the printed label on a plant and seen two names, perhaps Cornflower followed by *Centaurea cyanus*. What? What's *Centaurea cyanus* when it's at home? Well, Centaurea is the centaur's plant, from *Kentauros* in Greek, a mythical half-man, half-horse, who galloped into the ancient Greek imagination from mountain pastures in Thessaly. When centaurs were feeling poorly—claimed an old Greek folk story—they nibbled on cornflowers and were restored to studly vigour. *Cyanus* is the Botanical Latin form of *kyanos*, a Greek colour adjective meaning 'dark-blue' and here referring to one colour of cornflowers. Cyan is also a modern English colour word for 'dark-blue.' The same verbal root appears in the word *cyanide*. The common name recalls the fact that cornflowers grow wild in the grain fields of southern Europe.

The names we have given plants have interesting stories to tell. Both the common and the botanical names are easier to

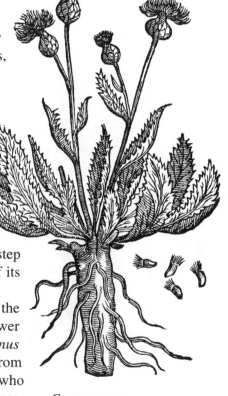

Centaurea or cornflower

remember if you know why a plant was named. This book demystifies botanical nomenclature. It is true that the scientific names for plants can look and sound like ungainly verbal monstrosities, the kind of words that, if they wanted a quick swig, would drop into that monster bar in the Star Wars trilogy. But in fact even long, complicated-looking words in botany have down-home roots. The source of these technical terms is often found in homey metaphors, uses of the plant in old folk medicine, mythology, and even old wives' tales. Discovering the reason for a plant name leads to knowledge of Latin and Greek roots which helps in understanding other words in English. Deconstructing what looks like learned gibberish is fun and can increase your vocabulary.

Botanical names need defeat no gardener. Knowing these scientific labels gives a gardener trying out a new plant some knowledge of that plant's native locale or habit of growth, and so the gardener has a better chance of growing it successfully.

A centaur subdued

And of course this book seeks to answer the important questions in Canadian life. Why was spruce tea once peddled as "The Great Antiscorbutic Elixir of the Canadas, Most Easily Made and Necessary To Be Had by All Persons There Resident"? What tree did both Canadian beavers and the builders of Venice use? A snotty var is a familiar

Canadian tree in Newfoundland. Which one? Begonia is named after a governor of French Canada. Who was Michel Bégon? When Acadians were driven out of Nova Scotia, some settled in Louisiana to become Cajuns, and found down south a tree that also grew in their Acadian homeland. In fact, they named the capital of Louisiana after this tree, Baton Rouge or 'red stick' in French. But which tree reminded them of the home from which they had been so ruthlessly expelled? Which human body part is named after a pine cone? Where did we get our labels for plants? Which plant names are Canadian in origin? The introduction gives a hint.

For their assistance beyond the call of commerce, I thank publisher Kim McArthur and publicist Laura Cameron, and all the enthusiastic crew at Little, Brown (Canada). Cheerful dynamo Kim McArthur's support for this book included handholding, brow-wiping, and quelling of authorial exasperations. Once again my editor, Pamela Erlichman, has pruned the deadwood from my prose, and designer Tania Craan has presented it in a typographic garden of delight. My friend Keith Thomas has kindly noted mistakes in general scientific outlook, and corrected them with all the warmth and self-effacement of Captain Nemo impaling live bait on a fish hook. Darren Hagan helped me to index the book and to keep the rows of this paper garden straight, even the ones that were hard to hoe. I also thank my agent, Daphne Hart, of the Helen Heller agency, for her contractual dexterity and friendship.

An old rural saying from the Canadian Prairies states that "the best fertilizer for the soil is the farmer's footprints." So pull on those old garden shoes. I know you'll enjoy the tromp. Just follow me. Any questions or additions? Drop me a note.

September, 1997
205 Helena Street,
Dunnville, Ontario, Canada N1A 2S6

COMMON CANADIAN PLANT NAMES VERSUS BOTANICAL BINOMIALS

Canadians have names for the plants of the field, of the mountain, of the sea, and, perhaps best loved, of the garden. Camas bulbs are grown all over the temperate parts of the world. The plant and the word are Canadian in origin. Camas, camass, commas, kamass, or quamash—there are more spellings for this once staple food bulb than you can shake a digging stick at. Its botanical tag is *Camassia quamash*. Once Pacific coast peoples used to harvest the bulbs of this blue-flowered member of the lily family and bake them immediately in ground ovens. They could be eaten hot or dried and stored for winter rations. People harvested them in the plump-bulbed autumn.

Camas is the Chinook Jargon descriptive which came from the Nootka adjective *kamas* 'sweet.' The original Nootka name for the place that became Victoria on Vancouver Island was *Camosun*, 'place where we gather camas.'

From seaweeds like New Brunswick's dulse, to southern Alberta's prickly pear cactus, we have also borrowed plant names from other languages and peoples. The green tapestry of our common and botanical plant names is a story many gardeners have neglected. And Wee Willie of the Greenwood, *moi*, well, shucks folks, I'm here to set that right. Did you know early explorers of the Canadian West ate a little delicacy called rock tripe? Was it the "guts" of a rock? Voyageurs thought so, and were grateful to make a sticky gumbo-like stew out of a lichen found on rockfaces along the Fraser and the Kootenay Rivers.

Jack-in-the-pulpit, *Arisaema triphyllum*, the familiar little arum of our moist woods, takes its generic name from Greek for 'blood arum' because certain European species have reddish

SURPRISING PLANT NAMES

pansy = *pensée*, the French word for thought, because the flowers nodded sagaciously.

cowslip = *cu slyppe*, which is Old English for 'cow dung.' The plant grows well in pastures and mucky places.

nasturtium = 'nose-twister' in Latin. Its flowers smell pungently.

spots on the leaves. French-Canadian lumberjacks encountered the plant in Québécois woodlands and thought the spathe that bends over the "jack" or spadix like a little pointed flap looked more like one of their implements, a hand-held log-hook or *gouet*, hence the North American French name, *gouet à trois feuilles* 'hook with three leaves.' Iroquois peoples named it *kaha-hoosa* 'papoose cradle,' because it looked like the backrack Iroquois women used to carry their babies.

But we also have surprising sources here of plants not Canadian in origin but known to every Canuck. Jingle your bell on mistletoe.

Are all plant names sweet and gentle, summoning to mind Little Mary Green Thumb skipping merrily down her garden path in a gingham frock and Victorian sun hat? No indeed!

MISTLETOE

It is so Christmassy! But in its original Old English form, *misteltan,* mistletoe means 'shit-on-a-stick." Oops! The ancient druids held mistletoe sacred, hence an old common name of mistletoe, druid's herb. The druids thought the plant grew miraculously from bird droppings. In Old High German, *Mist* meant dung. English belongs to the West Germanic branch of Indo-European languages. *Mistel* was the name of the white mistletoe berries in Old English. *Mistel* meant literally "shitling" or "little dropping" and *tan* was one Old English word for "twig" or "stick."

Long after the druids, botanists discovered that mistletoe can be propagated from feces dropped on upper tree branches by birds, after they have eaten the seed-filled berries.

Think not of this, when next you kiss under the mistletoe.

Orchid means 'testicle' in ancient Greek. The flower that red-faced dudes pin to maidenly bosoms on prom nights was named after certain species of ancient Attica that had twin roots resembling the human scrotum. Does macho claiming then lurk in this innocent dating ritual where the male fastens a symbolic scrotum to his female partner? One hopes not.

Avocado is an early mangling by Spanish *conquistadores* of *ahucatl*, the word for 'testicle' in Nahuatl, an Aztecan language of Mexico and Central America. Vanilla flavouring is extracted from seed pods of an orchid first called *vainilla* in Spanish. *Vainilla* means 'little vagina,' named because the shape of the vanilla pods reminded a botanizing but lusty explorer of what he missed most. *Vaina* from the Latin word *vagina* is still in modern Spanish where it means 'sheath for a sword,' its prime meaning in Latin.

Consider too the plump purple of the noble aubergine, the British and European name for what we North Americans call eggplant—a dowdy, frumpish name. Aubergine goes all the way back to Sanskrit, a classical language of India, where it was called *vatinganah* which means literally 'fart, go away.' From Sanskrit, the word and the fruit were borrowed into Persian as *badingan*, then into Arabic as *al-badhinjan*.

When the Arabic-speaking Moors first conquered Spain they brought eggplants with them. Some Spaniards borrowing the term thought the Arabic definite article *al* was part of the word, so it went into the Catalan language as *alberginia*. The French picked this up as *aubergine*, and English nabbed the word late in the 1700s. As to the anti-flatulent property of eggplants, eaters vote on both sides of the question. Some say aubergines reduce intestinal gas; others claim their consumption is an invitation to a tooting patootie.

The daisy, that perky little upstart of side-beds and ditches, began life in Old English, long before the twelfth century, as *dæges eage* 'day's eye.' The letter *g* between vowels in Old English had a soft *y* sound. The eye of day—splendid tag for a daisy!

As I hope you begin to see, knowing the origins of common plant and flower names adds to the joy of gardening. But this book celebrates all plant names, common and botanical. Every language on earth has a stock of simple verbal labels applied to the several hundred plants most important to the speakers of that language. Plants that receive common names are useful for food, shelter, medicine, and ornament. Harmful and poisonous plants also tend to get named—after all, we have survived since our dawn on the African savannahs. Plants of no obvious and immediate use to humans had to await formal botanical nomenclature, which began with the Swedish founder of plant taxonomy, the naturalist Linnaeus, in 1753.

Some green-tower botanists blather away against the use of common names and loudly bray that everyone ought to use botanical binomials. Nonsense, of course. Such "experts" put the monotony in botany. This book, I trust, removes it.

Any ramble in the garden of names leads to the gate of this conclusion: sufficient to their daily uses are the common names of plants. Garden-variety labels suit the hundred or so plants most gardeners and farmers encounter in a life of nurturing ornamentals and crops.

Canadian Garden Words boasts sections about the word lore of annuals, bulbs, herbs, perennials, trees, and wild plants native to Canada. But, if you need a brief refresher on botanical names, please browse the next section.

A QUICK OVERVIEW OF BOTANICAL NAMES

The science of botany classifies plants by a system with many categories. All living things that are not animals are grouped in a vast and capacious kingdom called Plantae, the plants. In this kingdom are categories from the largest to the smallest groupings: divisions, classes, orders, families, genera, species, and varieties. This book concerns especially the names of family, genus, and species. It seeks to make the botanical name of a species, as well as its genus and family names, understandable, by showing with clarity and a bit of humour, how and why the plants we grow in our gardens and see in our Canadian landscape received their formal labels.

The scientific names of plants are recorded in Botanical Latin, a special form of the ancient language spoken and written by the Romans. For many hundreds of years, scholars who studied our green world used Latin and Greek words to make up the names of plants. And they still do so, chiefly because a name in Latin allows the same plant to be referred to all over the world, by a unique ID tag that is understood by a botanist or gardener who speaks Turkish, Hawaiian, Russian, French, or any other language. Since one plant will have many, different common names in all these languages, using a unique Latin name for each species of plant assures scientific clarity and precision in identification. One, simple, internationally accepted, Latin label saves having to translate the plant's name into individual languages.

Carefree author demonstrates leisure activity almost as enjoyable as memorizing botanical names.

"Then some dude wearing a white wool blanket tells me that Wotan's real name is Jupiter!"

But why do botanists use Latin and Greek roots to form their technical names? The simple answer is: 90 percent of all scientific words in English derive from Latin and Greek.

In origin, English is a Germanic tongue based on the Germanic dialects of the Angles, Saxons, and Jutes who conquered Britain. But invasions of and migrations to the British Isles of peoples speaking other languages, like Romans speaking Latin, Vikings speaking Old Scandinavian, and Normans speaking French, added foreign terms to the basic Anglo-Saxon word hoard. Consequently, English now has more lexical items, more words, than any other language on earth, a larger vocabulary than Chinese or French or Russian or Arabic. True, our simple words are still of Anglo-Saxon origin: 'give,' 'man,' 'father.' But almost all our technical and learned words have been borrowed, sometimes through Norman French, from Latin and Greek. For example, *donate* is a Latin-based verb meaning 'to give,' *human* is a Latin-based word referring to mankind, *paternal* is a Latin-based adjective referring to a father. Latin terms were borrowed earlier and are often more familiar than Greek terms borrowed later. For example, *didonai* is the Greek verb 'to give,' from which English gets antidote, something given against a toxin, to counteract its effects, from *anti* Greek, against + *dotos* Greek, given. Anthropocentric is a relatively recent coinage, based on *anthropos* Greek, man + *kentrikos* Greek, centred. Patriarchy is rule by older males, literally by fathers, from *pater*, *patros* Greek, father + *arche* Greek, supreme power, dominion, rule.

Why Latin? For almost two thousand years, up to the end of the seventeenth century, scientific textbooks were written in Latin. If you were a student at the Sorbonne, or at Oxford, or Bologna, you learned natural history (later called botany and zoology) from books written in Latin but based largely on the writings of early Greek scientists.

But again, why use "dead" languages, Latin and Classical Greek, to form scientific and technical terms? First, it is traditional—as we saw above. Second, in a "dead" language, the meaning of a word does not change. It is frozen. *Callus* will always mean 'hard skin' in Latin. In a living language, words acquire new meanings. In 1930, acid meant a chemical like the acetic acid in vinegar. Nowadays "acid" is English slang for LSD, a dangerous hallucinogenic drug.

Because precise meaning and precise use of words is crucial in all forms of scientific communication, it helps to be able to make new botanical terms from Latin and Greek roots whose meaning does not alter over time.

As you read about the exotic origin of plant names, you will see that much Botanical Latin is derived from ancient Greek words. Why? First, the Greeks got around to studying and naming plants long before the Romans did. So there exists in ancient Greek texts a large vocabulary of plant names. Second, compared to the Latin language, ancient Greek simply had more words, had a larger and more sophisticated vocabulary. Latin is a terse tongue, a language that valued concise utterance. Thus Latin has few words with many meanings. In Latin, context is everything. This is not as true in Greek, a language with an inherent predilection for forming compound words with felicity, to produce pleasant-sounding and logical names. Unfortunately this aptness and euphony of nomenclature does not hold for all botanical names formed from Greek roots by modern botanists. Some of these new terms are frankly ugly and incapable of being pronounced easily. Yes, there are compound words in Latin, but not nearly as many as in ancient Greek. Stated plainly, it was easier to make new words in Greek than in Latin. Now, let's look at some categories of these plant names made from Latin and Greek roots, especially at the names given to an individual genus, species, variety, and family.

THE GENUS NAME

All of us who garden or grow plants commercially use some genus names in Botanical Latin every day and think nothing of it: Aster, Chrysanthemum, Delphinium, and Geranium. Aster is the name of a genus. A genus is a basic group of plants. Plants in one genus are more like one another than they are like any other group. Aster is called a generic name. Some other generic names are: Anemone, Crocus, Fuchsia, and Petunia.

THE SPECIES NAME

The second part of a botanical name identifies the species. It is called the species name or the specific name or the specific epithet. Epithet comes from the ancient Greek word for 'adjective.'

A species (from *species* Latin, appearance, a kind, a sort) is the most basic unit used in classifying and describing a living organism. Individual plants of the same species can usually interbreed and this possibility of exchanging genes is one of the reasons they are grouped together. An example of a specific name in the Aster genus is: *Aster novae-angliae*. Note that full botanical names are usually set in an italic typeface and only the genus name is capitalized. The specific part of this botanical name, *novae-angliae*, means 'of New England' and indicates that this species of aster is native to eastern North America including southern Canada.

Long ago, British and European plant breeders collected samples and seeds of *Aster novae-angliae* and used it to create garden varieties of asters. In England, these plants which bloom in the late summer and fall are called Michaelmas daisies. Michaelmas (St. Michael's Mass or Feast) is an Anglican and Roman Catholic religious celebration of Saint Michael which falls traditionally on September 29, and, usually, these asters are in bloom near that date.

But, how apt is this British common name? Michaelmas daisy. Well, it is an aster, not precisely a daisy, although some daisies reside in the genus *Aster*. However, hundreds of different species are commonly called daisies, many not even in the daisy family of Compositae. Many varieties of this aster bloom at the end of August, long before Michaelmas. You decide which name is more apt. *Aster* is the Latin word for 'star' and refers to the shape of the flowers in this genus, to the star-like, radiating arrangement of its petals.

ASTER

The Latin and Greek word for star gives many English words.

Disaster occurs under an unlucky star.

An asterisk * is a star-like symbol used to mark off important words. It comes from a Greek word 'asteriskos' that means 'little star.'

An astronaut is literally in Greek 'one who sails to the stars.'

CULTIVAR AND VARIETAL NAMES

There is often a third part to a botanical name, in which the particular variety or cultivar or subspecies of a plant is identified. A cultivar is a variety of a plant produced by selective breeding to perform well as a garden subject. Let's create a cultivar name out of the blue: *Aster novae-angliae* var. *Alberta Sunset*. This is—let us suppose—a wonderful, big aster that can grow to five feet with flowers coloured a rich salmon-pink, very much like some sunsets in Alberta. This third part of a botanical name may be called a varietal epithet or a cultivar name or a subspecific.

FAMILY NAMES

Genera and species are grouped into larger units called families. By convention, most names of botanical families end in the arbitrarily chosen termination *-aceae*. For example, the Rose

Lilium flo. Alb.   Lilium bisantinum.

family is Rosaceae. The Lily family is Liliaceae. The Iris family is Iridaceae. A few families that were named very early in the history of botanical nomenclature have older-style familial labels. For example, the Daisy family is Compositae, the name derived from *compositus*, a Latin adjective that means 'placed together, compound.' This largest family of flowering plants is named after the compound flowers of its members. Small florets of individual flowers make up large clusters or heads. Other older-style family names are Labiatae, the mint family; Leguminosae, the pea family; and Umbelliferae, the carrot family.

And that's really all there is to an initial understanding of botanical names. If you know why an aster is called an aster, you will remember the scientific or botanical name more easily. Communication about, reading about, and researching plants, all these activities become easier too. And that is why I wrote this book. It is good to know the common names of plants, but gardening is more interesting if you can use and understand the botanical names as well.

THE SHOCKING SHAMROCK EXPERIMENT

In 1991, botanist E. Charles Nelson wrote *Shamrock*, a definitive book about the plant long symbolic of Ireland. As research, Nelson repeated a famous botanical experiment designed to test the validity of common names for plants. He sought to determine the exact scientific identity of the true shamrock. On or about one St. Patrick's Day he received 221 plants from thirty Irish counties, all shamrocks in the minds of those who sent him the plants. But there were four different plants the Irish called shamrock! White clover (*Trifolium repens* or literally in Latin 'creeping three-leaf'), red clover (*Trifolium pratense* or 'meadow three-leaf'), lesser yellow trefoil (*Trifolium dubium*), and spotted medick (*Medicago arabica*). If he had canvassed England, he would have received a fifth plant, wood-sorrel (*Oxalis acetosella*). Common names are appropriate but with this caution. After they have been in use over a wide area for many centuries, common names tend to be applied to a larger and larger number of different plants. If one relied only on local names for plants, much confusion would follow. The marsh marigold whose yellow clusters brighten swampy areas of the temperate zone, has more than 80 local names in Britain, over 60 in France, and 140 in Germany!

Clearly, there is a need for botanical names. Common and botanical names are both necessary, both interesting, both worth studying—as I hope the following pages prove. Our table of contents provides a useful survey of the plan of this book. Use it to focus on groups of plants that interest you.

And—happy gardening!

ABBREVIATIONS & LINGUISTIC TERMS

Dianthus

1. What do < and > mean in linguistic descriptions?

Linguistics has borrowed these two arithmetical operator signs. As a mathematical symbol > means 'greater than' and < means 'less than.' But in language study, > means 'develops into, is the root of.' And < means 'stems from, derives from, is related to.'

2. The asterisk * and Indo-European

The asterisk (*) indicates a hypothetical form, usually in the Indo-European (or IE) family of languages to which English, and most other languages of Europe belong. Consult a large dictionary with a chart of the hundreds of languages which belong to this ancient family. The forms are hypothetical and marked with an asterisk because the Indo-Europeans had no system of writing, so the prehistoric roots are adduced from their modern forms. This ancient mother tongue is sometimes called PIE for Proto-Indo-European. It accounts for the similarity in many common terms across now widely separated languages, for example, explaining why father, *pater*, *pitar*,and *Vader* are all related and all mean 'father.'

3. Why are there so many forms of some Latin and Greek words?

The simple or nominative case of a Latin or Greek noun is not always the form used to make a Botanical Latin term. For example, many Latin and Greek nouns do not contain the full root form in the nominative case, but do show it the genitive. Thus, it is traditional in learning Latin and Greek nouns to quote a noun in its nominative and genitive case. You have seen it frequently in dictionaries. You will see it here occasionally. For example, one might list a root like *chrom-* and say it derives from *chroma, chromatos* Greek, colour of the skin, then any colour; used as a combining form signifying 'colour.' The Greek is given as *chroma, chromatos* because English has borrowed both the nominative and the following genitive of the Greek word for colour to make new scientific words. For example, in botany, the word *chromosome* uses only the nominative stem. And the English scientific word *chroma* uses the Greek nominative in its original form. But the adjective referring to colour, chromatic, uses the genitive stem.

Like all languages whose nouns have declensions of different case endings, and whose verbs have many conjugational endings, Latin and Greek words had many forms. Different parts of Latin and Greek nouns and verbs have been used over the centuries to make new botanical names. That is why I sometimes show a classical word in several forms.

ANNUALS

HOW & WHY THIS PARTICULAR ANNUAL LIST GREW

Listed alphabetically by common name, these familiar annuals are generally available from Canadian seed companies and nurseries. If the botanical name is more widely known than one of the common names, then the annual has been listed by its botanical name, for example, ageratum, impatiens, nasturtium, phlox.

I tested my little lists on gardeners in various provinces of Canada to see if they recognized so-called "common" names. What I learned about species recognition among people who are far better gardeners than I am was a bracing corrective to those garden guidebooks that list thousands of annuals and perennials by thousands of common names. In my unscientific and amateur test of Canadians who garden, there was general recognition by common name of about twenty-five annuals and perhaps fifty perennials. Some plants have literally a dozen common names! So how was I to list, for example, *Amaranthus caudatus*? Its common name Love-Lies-Bleeding was recognized by gardeners on southern Vancouver Island, in Alberta, Manitoba, southern Ontario, and Nova Scotia. Yet some friends in Halifax called it Tassel Flower. It was also identified with other common names applied to other species of amaranthus, namely, Prince's Feather, Joseph's-Coat, and Fountain Plant. My listing rule became: alphabetize under common name only if common name enjoys wide recognition; otherwise list under botanical name.

Searchers for the word lore surrounding a particular plant should remember to consult the extensive index, where common names and botanical names are all listed separately.

One must balance the vision of what a book might be with what it can be. From the specialist who finds an exotic seed for sale in a foreign catalogue to the gardener like me who buys flats of alyssum and petunia every spring for bedding, North Americans grow thousands of plant species, so consideration of book length made me prune the initial list, favouring common

An amaranthus from a 17th-century woodcut.

annuals dubbed with names of interesting origin. Sure, I wanted a 5,000-page multivolume extravaganza bound in unborn pterodactyl leather with gold-leaf titles applied by tooth by armless monks on a small island near Malta. But, kindly pragmatist that she is, my publisher demurred, as did ecology mavens and toothless monks.

Let me add a thought embued with personal poignancy. A friend once nicknamed me *Bibliosaurus casselmanicus*, conjuring some wheezing geezer of Mesozoic provenance. While not that old, I can murmur as Horace did, "Alas, the fleeting years are flowing by." I don't have the half-century required to compile some colossus of horticultural inclusivity. All listing is selective. My wish is that, as a reader, you will find my pruning attractive and engaging.

Special consideration was given to plant name origins neglected in gardening encyclopedias and to spurious bits of folk etymology or plainly wrong origins passed down in dictionaries and botany textbooks— sometimes for centuries! But the chief criterion was delight in word lore. I hope it will intrigue readers to know that wax begonia, compact borderer of garden beds, may honour the memory of Michel Bégon, a man who as *Intendant* of New France once controlled part of what became Canada.

Unfortunately, every plant sold by the garden trade as an annual is not an annual. The categories in this book are everyday, garden-variety categories, used for ease of reference. They are not necessarily scientific groupings. Some biennials and perennials will bloom from seed in one year, as all true annuals do. But for practical trade and garden use, we follow custom and class them as annuals.

Dianthus

Common names: Floss Flower, Pussyfoot

AGERATUM

Genus: Ageratum, Latin < *ageratos* Greek, not growing old < *a* Greek, not + *geras* Greek, old age, referring to the fact that one of the first collected specimens kept its lavender-blue flower colour well. Linnaeus named the genus.

Family: Compositae, the daisy family < *compositus* Latin, placed together. This largest family of flowering plants is named after the compound flowers of its members. Small florets of individual flowers make up large clusters or heads. The standard example of Compositae is the daisy.

French: *agérate*

Word Lore
Ageratum shares the *ger* root with words that concern the scientific study and treatment of ageing: geriatrics, gerontology, and a disease called progeria caused by genetic defects that produce premature old age in the very young.

Species
The ageratums of commerce are hybrids of Mexican and Central American species, hence the two main sources of nursery stock, *Ageratum mexicanum* and *Ageratum houstonianum*. William Houston (1695-1733), a Scottish doctor, collected a few of the thirty species in Central America and Mexico and introduced them to British seedsmen.

Ageratum

❧

Common names: Sweet Alyssum, Snowdrift, Madwort

ALYSSUM

Genus and Species: *Lobularia maritima* < *lobulus* Latin, little pod, refers to the fruit; the specific is maritime because Sweet Alyssum is native to the northern shores of the Mediterranean Sea.

Family: Cruciferae, the mustard family < *crux, crucis* Latin, cross + *ferre* Latin, to carry, to bear. The cross they bear is the arrangement of their flower petals, usually four petals at right angles to one another, making the grouping resemble a little cross. A large family, the crucifers include many important vegetables; some like cauliflower and broccoli are considered anticarcinogenic, others are nutritious like cabbage, brussels spouts, turnip, and radish.

French: *alysson, alysse*

WORD LORE

Alyssum
Latin < *a* Greek, not + *lyssa* Greek, madness, rage, literally: a loosening of the mind. Ancient herbalists thought this plant cured some forms of madness. It was falsely thought to help cure rabies contracted from the bite of mad dogs. *Alyssum maritimum* is the old botanical name of this annual favourite of rock gardens and borders, whose honey scent wafted on a warm summer breeze makes it attractive to bees and pleasant to plant beside a veranda or patio.

Madwort
The old English name, madwort, is a partial loan-translation of alyssum as 'mad plant.' Wort < *wyrt* Old English, root, herb, plant is cognate with Old Norse *rot* which was borrowed early to give *root*. Wort is also cognate with modern German *Wurzel* 'root' and *die Würze* 'spice, seasoning' and *das Gewürz* 'spice, herb' and with the German varietal grape and wine, *Gewürztraminer*, literally 'spicy-tasting (grape) of Tramin,' the German version of the Italian place name, Termeno. Wort produced many compound plant names in early English. Some are still in use, others obsolete. Consider bellwort, bitterwort, bladderwort, figwort, liverwort, lousewort, lustwort, moonwort, sneezewort, and spearwort.

Also related are *radix, radicis* Latin, root, and *rhiza* Greek, root, both of which give horticultural words in modern English like radish, eradicate 'to root out,' radicle, and rhizome. Even the political adjective *radical* begins in the garden at the dawn of the nineteenth century, when advocates of radical reform like Henry Hunt wished to reshape the Liberal Party of Great Britain "from the roots up." Later radical reformers said—redundantly—that they were "eradicating the roots of evil."

Licorice hides its Greek root *rhiza*, having suffered many a change as it came through Old French from Late Latin *liquiritia*, an illiterate attempt at Late Latin *glycyrrhiza* from the original Greek name, *glykyrrhiza* = *glykys* sweet + *rhiza* root. Licorice flavouring, used to tart up candy, liqueurs, and medicines, is an extract of the dried root of a European legume, *Glycyrrhiza glabra*.

AMARANTH

Common names of several species: Flaming Fountain, Joseph's Coat, Love Lies Bleeding, Molten Fire, Prince's Feather, and Tassel Flower

Genus: Amaranthus, Botanical Latin < *a* Greek, not + *marantos* Greek, fading. Many species have flowers and coloured leaves that retain their colour for two months. Unfading indeed! In the first century A.D. a Greek writer on the medical uses of plants named Dioscorides wrote of the *amaranton*, an unfading plant of similar habit, although he appears to refer to a member of Compositae, certainly not to the flashy amaranths of modern horticulture which are natives of the tropical Far East and were quite unknown to the ancient Greeks.

Family: Amaranthaceae. The *-aceae* suffix that terminates most names of botanical families is an arbitrary clump of pseudo-Latinate adjectival plural endings that makes the family names look and sound monstrous, but this ungainliness is enshrined in holy botanical writ, and so we shall deal with it, never forgetting it is ugly word-making at its unspeakable worst.

French: *amarante*

SPECIES

Prince's Feather

The most commonly grown amaranth is the floppy, weedy
Prince's Feather, *Amaranthus hypochondriacus*. Say that six
times with a mouthful of chewed peas and you'll have a new
design on your kitchen wall. Contrary to what is suggested by
the authors of *The Harrowsmith Annual Garden* (1990),
hypochondriacus does **not** mean the plant had "former medical
uses." In botany the specific epithet *hypochondriacus* always
refers to a plant of sad, mournful stance, sickly like a
hypochondriac. Prince's Feather is a gangly, droopy thing that
grows quickly and rankly and usually needs staking. Another
species sold as Prince's Feather is *Amaranthus hybridus
erythrostachys* (*erythros* Greek, red + *stachys* Greek, spike;
referring to the flower heads).

WORD LORE OF HYPOCHONDRIA

This Greek word is made up of *hypo* below + *chondros*
cartilage + *ia* diseased condition.
The ancient Greeks used this word to refer to the area of the
upper abdomen below and behind the cartilages of the lower
ribs. It is still sometimes called the hypochondriac region.
Ancient Greek physicians also believed that feelings of distress,
depression, and general anxiety about one's health came from
the hypochondriac region, and this is the origin of the term
hypochondriac for a person with chronic, unrealistic concerns
about his or her health.

Hypochondriasis is the preferred medical term for the older
word *hypochondria*. Hypochondriasis is used by some clinicians
for extreme patients who plunge into dangerous depression
convinced they have diseases that rational medical evidence
says they do not have.

Joseph's Coat

This splendid, two-metre-high giant, a waterfall of coloured
leaves, can stand proudly at the back of any border. *Amaranthus
tricolor* has one cultivar "Molten Fire" that looks like a sky
rocket of copper flame. Some gardeners use Joseph's Coat as a
temporary shrub, but its gaudy colours can drown out the more
modest hues of companion plants.

Love Lies Bleeding

Amaranthus caudatus (Botanical Latin, with a tail) has a species name that refers to the red flower spikes that droop like clustered tassels from the top of the plant against the red and green leaves.

~

Common names: Alkanet, Bugloss, Cape Forget-me-not

Genus: Anchusa, Latin < *agchousa*, *egchousa* Greek, a kind of rouge, and the plant from which this rouge and other dyes were made > *egchousizein* Greek, to put on rouge < *egchein* Greek, to pour on, to apply

Family: Boraginaceae, the borage family, a small group of European and Asian herbs grown for red dye from their roots and for their blue flowers which are still used to garnish salads. Borage leaves are used to flavour punches and cocktails. Borage < *bourrache* Old French < *borrago* Medieval Latin < *abu 'araq* Arabic, father of sweat, because early Arabic medicine used borage as a sudorific, an agent that causes sweating.

SPECIES

Anchusa azurea (Botanical Latin, deep blue) is the southern European perennial flower of deep, true blue whose roots were used to make a red dye.

   Anchusa capensis (Botanical Latin, of the Cape of Good Hope, South Africa) is the species used as a garden annual because of its skyblue flowers.

   Anchusa officinalis is the southern European anchusa most used historically to obtain blue dye from flowers, and red dye from roots. But the Herb & Spices Set, when in an arty mood, also float alkanet flowers in wine punches and fruit cups.

   Officinalis is an interesting medieval Latin adjective seen as the specific name of many older botanical plants that were sold in shops because they had medicinal and cosmetic uses. *Officina* is the medieval Latin word for a workshop or shop, from which

ANCHUSA

This and many of the subsequent illustrations are from 15th- and 16th-century woodcuts. The old botanical names have, in some cases, been revised and corrected by modern botanists.

English gets, of course, *office*. *Officina* is a contraction of *opificina* from *opifex* 'craftsman, mechanic,' which in turn is made up of *opus*, Latin, work + *fex, ficis* Latin noun and adj. suffix, doing, making. Compare *facere* Latin, to do, to make.

Officinalis as a specific in a plant name almost always indicates the plant was widely used in medieval times and often centuries before for some human purpose.

WORD LORE

Alkanet

A good example of the muddle of common plant names in Europe before Linnaeus regularized scientific plant names, alkanet has been applied through history to five utterly different plants, whose only common feature was that they all supplied some kind of dye from their parts. Alkanet today is still used as a common name for several Anchusa species. The Moorish Arab conquerors of Spain brought the word to Europe as *al-hannat*. Does that look like henna? Well, it is. Henna is a different plant, a Middle Eastern shrub, *Lawsonia inermis*, whose powdered leaves give a reddish-brown dye used for thousands of years to colour hair and cloth. Classical Arabic is *al-hinna*; Colloquial Arabic is *al-hannat*. The *t* is just the common Semitic marker for nouns of feminine gender. Spanish turned it into *alkaneta* and applied it to native Anchusa species, not to true henna, and then the word made its way into all the languages of medieval Europe, and into English as alkanet.

Bugloss

Bugloss is an old common name for several members of the borage family, like *Anchusa officinalis* and *Anchusa arvenis* (Botanical Latin, of fields < *arvum* Latin, ploughed land < *arare* to plough, to till). Anchusa species were called this because their blue flowers reminded someone of the bluish tongues of some domestic cattle. Bugloss < *buglossus* Late Latin < *bouglossos* Greek, ox-tongued < *bous* Greek, ox + *glossa* Greek, tongue. That Indo-European root for ox or cow sneaks into Old English cow-calls like "Co', Boss!" (Come, cow) and hence to common bovine farm names like Bossie.

Uses
Ancient Greek women made a blue eye-shadow from the
flowers, and a rouge from the roots. John Gerard, Elizabethan
gardener and author of the famous *Herball* of 1597, wrote of
anchusa that "the Gentlewomen of France do paint their faces
with these roots."

∾

**BABY'S
BREATH**

Genus: Gypsophila < *gypsos* Greek, lime, gypsum + *philos*
Greek, loving, lover; some species grow in limey soils.

Family: Caryophyllaceae, the pink family, named after its
typical species, *Dianthus caryophyllus*, clove pink. *Karya*
Greek, walnut + *phyllon* Greek leaf + *aceae* familial suffix of
botany. Walnut leaves have a sharp aroma; so do the leaves of
clove pinks. The family has at least seventy-five genera and
twelve hundred species.

French: *gypsophile*

SPECIES
Gypsophila elegans (Latin, selected, splendid, outstanding,
elegant), annual Baby's Breath, is native to the Caucasus and
western Asia. The airy panicles of tiny white flowers are
favourite cuttings of bouquet-makers.

∾

**BEGONIA,
FIBROUS**

Other name: Wax Begonia

Genus: Begonia < named after Michel Bégon (see below)
+ *ia* Latin noun suffix used frequently in botany to make plant
names based on surnames like Lobelia, Rudbeckia, and Zinnia.
The begonia genus has almost 900 species, most tropical, some
like Rex begonias grown for their beautifully coloured leaves,
others like tuberous begonias for their brilliant red, orange, and
yellow flowers. Fibrous begonias are used as annual bedding
plants.

Family: Begoniaceae, the begonia family

French: *bégonia*

SPECIES

The Wax Begonia, hybridized into dozens of horticultural forms, is *Begonia semperflorens* (*semper* Latin, always + *florens*, Latin, flowering). Widely used as annuals, Wax Begonias flower early and continue blooming outdoors until cut down by frost. These little natives of Brazil bloom in shade or sun as long as their requirement of a rich, moist soil is met.

WORD LORE

The Case of the Two Michel Bégons

The word *bégonia* was coined by the great French botanist Charles Plumier (1646-1704). Plumier was a French Franciscan monk who sailed to the West Indies to make botanical drawings of tropical flowers and ferns, exploring Martinique, Guadeloupe, and Haiti in 1689-90. Plumier brought back into favour an old practice of naming a newly discovered genus after people, hence his *bégonia*, instead of some montrosity like **Guadeloupia*. Although, as you will already know or soon discover in this book, botany is quite capable of forming the ugliest, jaw-breakingest, tongue-twistingest mouthfuls of pseudo-learned cacophony in its quest for new names. But, in defence of botany's many ungainly binomials, one must state that there are hundreds of thousands of named plants, and just as many unnamed wee denizens of earth waiting to be christened with a scientific appellation. Because the supply of Latin and Greek words applicable to botany is finite, name-seekers do get desperate from time to time. Linnaeus, who first systematized plant-naming, favoured Latin and Greek binomials for plants, although he slipped a few eponyms based on the names of friends and fellow botanists into his nomenclature too.

Humble monk Plumier wished to honour a governor of Santo Domingo who helped finance his island-hopping and botanizing. That person was Michel Bégon. Confusingly, there appear to have been two French colonial leaders with this same name. Was this plant named after Michel Bégon de la Picardière, an Intendant of New France who arrived to find the beaver trade of the colony in a depression? Bégon tried to reform the fur trade and start new industries other than pelt-sales that might eliminate the colony's deficit. He failed and was accused of various frauds and sent back to France in disgrace in 1723. Bégon's dates are given as 1667-1747. Some authorities list a Michel Bégon with dates 1638-1710, calling him a Governor of French Canada and a patron of botany. Others say he was a French governor of Santo Domingo who doled out a plump sack of francs to pay for Plumier's tropical sojourn, and then paid for specimens of Caribbean plants to be shipped back to Paris. Take your pick of Bégons and begonias.

∾

Common names: Tickseed, Beggar's Ticks, Beggar's Lice, Stick-Tight, Burr-Marigold

Genus: Bidens < *bi*, *bis* Latin, two, twice + *dens*, *dentis* Latin, tooth. Two-toothed refers to the two little prongs with barbs on the flat fruit or achene.

Family: Compositae, the daisy family < *compositus* Latin, placed together. The largest family of flowering plants is named after the compound flowers of its members. Small florets of individual flowers make up large clusters or heads.

BIDENS

SPECIES
Most species of Bidens are weeds. But *Bidens ferulaefolia* is a Mexican annual often sold as a basket-flower whose deep yellow flowers will bloom all summer as they cascade over the rims of a basket. Its specific *ferulaefolia* is Botanical Latin that means 'with leaves resembling giant fennel.' *Folium*, the Latin word for 'leaf' is the ultimate source of many current English words like portfolio, something to carry the leaves of a manuscript. Folio is shortened from portfolio. Gold foil is gold leaf. Cinquefoil has five leaves, in its original French, *cinque feuilles*. Trefoils in heraldry are three-leaved plant emblems. Forests were defoliated in Vietnam. Skin or bark can exfoliate, peel off in leaf-like sheets.

Ferula is the genus name of giant fennel which has tall, stick-like stalks that reminded an early botanist of the ferula.

*Bidens or
Bur-Marigold*

Ferula is a word invented by ancient Roman schoolboys. It means 'the little iron' and was the metal rod used by teacher-slaves to beat unruly students. The ferula was also used by masters to beat slaves. English has *ferule* for a usually wooden stick formerly used to punish children.

Tickseed is one of the Bidens. Anyone who walks through autumn fields or jogs along weed-infested roadsides in Canada has pulled the pesky little barbed fruits from jeans and jackets.

BLACK-EYED SUSAN VINE

Other common name: Clock vine

Genus: Thunbergia < Carl Peter Thunberg, Swedish botanical author, student of Linnaeus, plant collector who visited Japan, South Africa, and from many ports-of-call returned with plant specimens, including several of this genus of about 100 species of Asian and African vines and shrubs.

Family: Acanthaceae, the acanthus or bear's britches family < *akanthos* Greek, spine, thorn < *ak-* Greek root, pointed part + *anthos* Greek, flower. The pointy, prickly part of a plant is the thorn or spine. The root *ak* appears in all languages of the Indo-European family, where it means something with a sharp point. For example, in some English words derived from Latin, the *ak-* stem gives rise to: acid, acetic, acne, acrid, acupuncture, acute, and exacerbate. Some English medical terms borrow the Greek word directly. An acanthocyte (*acanthos* thorn + *cytos* cell) is an irregular type of blood cell in certain blood deficiency diseases, where the red blood cells appear distorted by little rays of protoplasm projecting from their surfaces giving the cell a thorny look.

The Acanthus Leaves of Classical Architecture
Acanthus spinosus is a prickly-leaved, perennial plant native to the shores of the Mediterranean. In their most ornate order, Greek architects decorated the capitals of their Corinthian columns with rows of stylized acanthus leaves, which the ancient Greeks considered to be the most elegant of all leaves.

SPECIES

Thunbergia alata (Latin, winged, referring to the petioles, < *ala,* Latin, wing), Black-eyed Susan vine, is a perennial that will bloom from seed in its first year given a long growing season, and so seeds of this delightful vine with its white and orange flowers are started indoors and set out well after frost has fled northern gardens.

∾

BLOODLEAF

Genus: Iresine < *eiresione* Greek, a branch of olive or laurel wrapped with wool and hung with fruit used in ancient Greek fertility ceremonies honouring Apollo and the sun < *erion* Greek, wool. Iresine refers to the woolly seeds and flowers of the two species common in horticulture. The common name Bloodleaf refers to the dark-red foliage for which the plant is grown.

Family: Amaranthaceae < *amaranthus*, Latin < *amaranton* Greek< *a* not + *marantos* fading. Many species have flowers and coloured leaves that retain their colour for months.

SPECIES

The two species in horticulture are perennials grown as annuals for their leaf colour, and seldom bloom in cultivation. *Iresine herbstii* is named after Carl Gottlieb Herbst (1830-1904) who found them while collecting in Brazil where he was director of the Botanic Gardens in Rio de Janeiro.

   Iresine lindenii honours a Belgian nurseryman named Linden who introduced this thin-leaved red variety to the trade.

∾

BROWALLIA

Common Name: Sapphire Flower

Genus: Browallia. Johann Browall (1707-1755) was the Bishop of Abo in Sweden and an amatuer botanist. Linnaeus, born Karl Linn (1707-1778), was the Swedish father of plant nomenclature. He formalized the two-word naming system for plants, which still consists in its simplest instance of a one-word genus name followed by a one-word species name. Linnaeus named

Browallia in gratitude for Bishop Browall's spirited defence of Linnaeus in 1739 when Swedish Christians objected to the Linnaean system of plant classification based on their sexual parts. Although counting the stamens and pistils and detailing their arrangement is no longer the only way plants are classified, it was an early, useful, and scientific approach.

Unfortunately Linnaeus wrote of this system playfully in an age when the sour, frozen Christians of Sweden were in no mood for sexual metaphors. Editorials of that time against Linnaeus roared that this dirty old botanist saw sex everywhere, even in innocent flowers! Linnaeus did not help by writing about "floral nuptials." He described flower petals as "bridal beds which the Creator has so gloriously arranged, adorned with such noble bed-curtains, and perfumed with so many sweet scents, that the bride-groom may celebrate his nuptials with his bride with all the greater solemnity."

Family: Solanaceae < *solanum* Latin, the black nightshade plant. The nightshade family also includes such important food plants as potatoes, peppers, tomatoes, and eggplants. In this family are *nicotiana tabacum*, common tobacco, as well as ornamentals like petunia and schizanthus, and poisonous plants like deadly nightshade.

SPECIES

Browallia speciosa major first discovered in the jungles of Colombia has white and blue, trumpetty, petunia-like flowers. *Speciosa* means 'showy' in Latin. *Major* means 'larger,' indicating there is a smaller-flowered species not common in horticulture.

Browallia viscosa compacta has a Sapphire-blue bloom with a white thoat. *Viscosa* means 'sticky' in Latin, and refers to the first growth to sprout which exudes a gluey substance to protect the young shoots from certain insects.

Species As a Word

A number of words we use everyday in English like special, specific, and spice, all derive from the Latin noun *species*, whose prime meaning is 'a sight,' related to the Latin verb *speciare* 'to look at, to see.' Hence a developed meaning of *species* is form or shape. Thus *speciosus*, or its feminine nominative singular form *speciosa*, as in *Browallia speciosa*, is made up of the base root *species* + *osus, osa, osum*, a Latin adjective ending meaning 'full of, abounding in.' A plant that abounds in shape or form is showy and is labelled *speciosa*. An interesting cousin of Latin *species* is the Italian word for mirror, *specchio*, something in which to see one's own form or shape.

In medieval Latin, *specificus* was coined as an adjective that meant 'form-making, giving defining shape to, pertaining to a thing's own peculiar shape or essence.' All later extensions of *specific*'s meaning stem from this medieval use.

In Late Latin, *specialis* was coined as an opposite of *generalis*. This evolved into the English adjective *special* originally meaning 'having its own form, shape, or quality.'

Again in Late Latin, *species* came to mean 'special article of merchandise' and then the meaning was narrowed to refer only to aromatics and certain imported herbs. In Old French, the Latin *species* became *espice* and was borrowed into Middle English as *spice*. Compare the modern French *épice* 'spice' and a place where one bought spices and other foods, *épicerie*, at first meaning 'spicery,' then 'groceries,' then 'grocery store.'

A related word in botany and other sciences is specimen. In Latin, *specimen* was in its basic sense something by which a thing was known or recognized, hence in Latin, a pattern, a model, an example. After being borrowed into the vocabulary of European science in the seventeenth century, it began to develop some of its present meanings like 'an example of something from which its chief characteristics may be inferred' and then 'something selected as typical of its class or genus.'

∾

Other common names: Belvedere, Fire Bush, Summer Cypress. Belvedere was borrowed into English from Italy late in the sixteenth century to name a gazebo-like summer-house with a pleasant view. At the same time it named a plant often placed near such little lookouts, summer cypress. Belvedere = *bel, bello* Italian, beautiful + *vedere* Italian, to see.

BURNING BUSH

Genus: Bassia < named after the first botanist to describe one species, Ferdinando Bassi (1710-1774). The former generic was Kochia, still used in some greenhouses and nurseries.

Family: Chenopodiaceae, the goosefoot family < *chen* Greek, goose; *chen* is of echoic origin, imitating the goose's call and thus an ancient Greek version of honk + *podion* Greek, little foot + *aceae* Latin, regular ending for family names in botany. The leaves of some species look like little goose feet.

SPECIES

Bassia scoparia has a specific epithet that means 'like a Scotch Broom plant' referring to its general shape and the way it leafs. A *scoparius* was a Late Latin occupational name for a sweeper, a man who used a broom often made of one of the Mediterranean Broom species. *Scopae* in classical Latin were twigs and branches tied together as a broom. The most common variety sold in trade is *Bassia scoparia trichophylla* (Botanical Latin, with hair-like leaves < *thrix, trichos* Greek, hair + *phyllon* Greek, leaf). In summer it has masses of thin leaves of a fresh, light lime-green. In autumn, the leaves fire a bright red. Given a hot summer, Bassia can shoot from seed to hedgy bush in one season. Chinese artists make one of their calligraphy brushes from the branches and leaves of a Bassia species.

∾

CANDYTUFT

Genus: Iberis < *iberis* Greek, a kind of pepperwort probably found by ancient Greek sailors who touched shore on some part of Iberia, prehistoric name for Spain and Portugal, where several species of candytuft grow wild. The ancient Greeks said the European peninsula was named after the *iber*, an animal, but the old texts make no mention of what kind of animal it may have been. What *iber* may have been is our common friend, the folk etymology, in which an animal name is made up to account for an otherwise inexplicable place name.

Family: Cruciferae, the mustard family < *crux, crucis* Latin, cross + *ferre* Latin, to carry, to bear. The cross they bear is the arrangement of their flower petals, usually four petals at right angles to one another, making the grouping resemble a little cross.

SPECIES

Iberis umbellata (Botanical Latin, with flowers presented in an umbel, a cluster in which the individual flower stalks all arise from one point) is the annual candytuft, provider of pleasantly scented, long-lasting cut flowers for indoor bouquets. Umbels were named because the "spokes" of the individual flower stalks in the formation look like the wooden struts of ancient Roman umbrellas.

WORD LORE OF UMBEL

Umbel < *umbella* Latin, literally 'little shade,' hence sunscreen, sunshade < *umbra* Latin, shade. Another diminutive form in Late Latin was *umbrella*, which also meant a little portable shading device to ward off the rays of the sun. Burnt umber is an artist's shady pigment. Penumbra (*paene* Latin, almost + *umbra* shade) is a word coined by seventeenth century astronomers to name the "almost shade" of a larger shadow or umbra, especially the shadow cast by the moon or the earth during an eclipse. A slightly hidden appearance of *umbra* is in the wide-brimmed Spanish hat, *sombrero* < *sombrar* Spanish, to shade < *subumbrare* Late Latin, to put under (*sub*) the shade, hence to overshadow.

❧

Other name: Bellflower, Bluebells, Wild Harebell

Genus: Campanula, Late Latin, little bell < *campana* Late Latin, bell, referring to the shape of the flowers + *ula* Latin diminutive suffix, little

Family: Campanulaceae, the campanula family

CANTERBURY BELLS

WORD LORE

The word for bell in many languages is often echoic. It imitates the sound of bells. For example, our English word *bell* is related to Old English *bellan* 'to roar, to make a loud sound' which also gives us *bellow* and the *belling* of male deer in rut. This naming of a thing or action by imitating its sound, as in buzz, fizz, hiss, is called onomatopoeia (Greek, name-making). One theory about the origin of Late Latin *campana* 'bell' is that *cam-pong* is a sonic equivalent to our English *ding-dong*.

Campanula

Another theory says *campana* stems from Campania, a flat, fertile region of ancient Italy around Naples whose mines produced metals from which ancient Roman bells were cast. Whichever theory may be true, *campana* did give the Italian word for bell-tower used in English, *campanile*, often used to describe a bell-tower which is separated from its church. Campanology is the science and practice of bell-ringing. Campania derives from *campus*, a Latin word for field. In medieval Latin *campania* was generalized to mean 'any flat country' and then even further generalized when it was borrowed into early French as *campagne* to mean 'countryside.'

The Romans had an equivalent of our English ding-dong and its dialect variant, tingtang. The Romans wrote the sound of bells ringing as *tin tin*. That gave the classical Latin word for bell, *tintinnabulum*, literally 'little *tin tin*' and the source of a fancy English word for a ringing sound, tintinnabulation. Related to the Latin noise words *tin tin* are the English words *tink*, *tinker*, and *tinkle*.

SPECIES

Campanula medium was the Late Latin name for a bellflower from Media in western Asia. The country prompted S.J. Perelman's nifty pun: "One man's Mede is another man's Persian." *Medium* here is not the similar Latin adjective for 'middle.' Some of these biennials will blossom the first year if started indoors. Native to many parts of North America, Asia, and Europe is the wild harebell, *Campanula rotundifolia* (Botanical Latin, with rounded leaves), known in Scotland as bluebells.

Other common names: Castor Oil plant, *palma Christi. Palma Christi*, palm of the hand of Christ, is the common name of Ricinus in many Spanish-speaking countries, being the medieval Latin tag for the plant. Often the leaves have five lobes and thus vaguely resemble human hands.

The handsomely marked seeds are poisonous to all domestic animals and humans. Gardeners who insist on growing this cumbrous, toxic oaf of a plant, please note: eight seeds ingested are fatal; there is no antidote. Why not avoid being one of the ignorant North Americans who is killed every summer by this ugly, deadly, gangly, botanical monstrosity? The phytotoxin is ricin. When ingested, ricin causes muscle spasms, convulsions, renal shutdown, and then death. Ricin is also a teratogen (Greek, monster-maker). In pregnant women, ingestion of quantities small enough not to be fatal to the mother has been proven nevertheless to cause birth defects, truly horrifying congenital malformations. Could we not rid Canadian gardens of this plant?

Genus and species: *Ricinus communis* (Latin, common) is the gigantic African herb which is the only species in this genus, grown as a tender annual in northern countries like Canada. *Ricinus* is Latin for a tick or a bug. The Romans thought the seeds resembled a certain Italian beetle, so here is another of the half-dozen plants called tickseed in various languages at various times. Castor oil is processed from the seeds of tree-like tropical plants, and is not poisonous after processing.

Family: Euphorbiaceae, the spurge family, named after a typical genus Euphorbia, which took its name from Euphorbos, the herbalist and physician of King Juba II (25 B.C.-23 A.D.) of Mauretania, an ancient country of North Africa that included parts of present-day Algeria and Morocco. Juba was known to Julius Caesar, was educated at Rome, and was renowned as a scholar, as was his court doctor Euphorbos, especially for his knowledge of the medicinal properties of plants. Many euphorbiacious plants are toxic. It is a large family that includes

CASTOR BEAN PLANT

CAUTION: POISON

the poinsettia, the crown-of-thorns, snow-on-the-mountain, the rubber tree of Brazil, and the gaudy crotons that winter in the house and summer on the patio.

Spurge entered English from Old French *espurgier* 'to cleanse,' itself from Latin *expurgare* 'to purge, to clean out the bowels.' Many species of the Euphorbia genus exude a bitter, milky-white sap which, when it is not toxic, is a dangerous purgative.

∾

CHRYSANTHEMUM, ANNUAL

Common names: Mums, Feverfew (a perennial chrysanthemum grown as an annual)

Genus: Chrysanthemum, Latin < *chrysos* Greek, golden + *anthemon* Greek, flower, in reference to the two Mediterranean, originally yellow-flowered species listed below and known to the ancient Greeks and Romans.

Family: Compositae < *compositus* Latin, placed together. The largest family of flowering plants is named after the compound flowers of its members. Small florets of individual flowers make up large clusters or heads. The standard example of Compositae is the daisy.

SPECIES

Chrysanthemum coronarium (Botanical Latin, used in garlands and flower crowns < *corona* Latin, crown) is sometimes called the Crown Daisy. While a wreath of pure gold-leaf might have crowned a victorious general returning to Rome in triumph, lesser Roman mortals worthy of a garland often had a wreath of flowers placed on their heads. Sometimes guests at a Roman feast were given flowery wreaths to wear as they reclined during the long meal.

Chrysanthemum segetum (Latin, of cornfields < *seges*, *segetis* Latin, grain field, crop). Much as some of us are of two minds about dandelions, ancient Roman farmers thought this little mum a pesky weed in a cultivated field, but a handsome yellow flower nonetheless.

Feverfew has the old botanical name of *Chrysanthemum parthenium* (Botanical Latin < *parthenion* Greek, virgin's plant;

compare an Old English common name for the same plant, Flirtwort. Perhaps it was considered an aphrodisiac at one time?). Modern botany has reclassified Feverfew as a tansy, *Tanacetum parthenium*, but we shall discuss it here. It is one of several plants with the common name, Bachelor's Buttons. Feverfew is as apt an example of folk etymology as you will encounter, being a mangling of febrifuge (*febrifuga* Medieval Latin, plant or medicine that drives away fevers). The Elizabethan herbalist John Gerard states that its leaves bound on the wrists are powerful "against the ague." For more than a thousand years, European folk remedies against coughing, wheezing, insect bites, colic, wind, and other maladies have often included teas and decoctions of feverfew leaves. It was frequently part of the herbalist's spring tonic.

❧

CLARKIA

Other names: Godetia (discontinued botanical name), Rocky Mountain garland, Farewell-to-Spring, Summer's Darling

Genus: Clarkia, named after the American explorer William Clark (1770-1838) on whose expedition with Meriwether Lewis to cross the American continent between 1804 and 1806 several species were first collected.

Family: Onagraceae, the Evening Primrose family < Onagra, former botanical name of the family < *onagra* Greek, a female wild ass < *onagros* Greek, male wild ass = *onos* Greek, donkey, ass + *agrios* Greek, wild; but why the name? Because some wandering botanist of yore saw wild asses eating evening primroses? We shall apparently never know. But early botanists did like the sound of the Greek word for donkey. Consider the Scotch Thistle whose botanical name is Onopordum from *onopordon* Greek, donkey's fart (*onos* donkey + *porde* fart). Did chomping on thistle make donkeys flatulent? Who did the field observations? The ancient Greeks had a clay jar with big handles like donkey's ears that they also called *onos*. The most important genus of Onagraceae in commercial horticulture is Fuchsia.

SPECIES

The wild species of Clarkia that grow in southern British Columbia at the northern limit of their range are surely Canada's prettiest wild flowers. *Clarkia amoena* (Latin, charming, pleasing) opens its cheerfully coloured little cup-like blossoms all across southern Vancouver Island, and can be seen less easily on mainland B.C. from the coast to the Cascade Mountains. In nature, Clarkia prefers cool summers and dry, sandy loam in its western habitat. It makes a long-lasting cut flower to beautify a porch table on a summer afternoon. Hybridists and makers of cultivars have tarted the wild species up, and seed catalogues often offer only double-flowered varieties of the North American species. But try to find a dealer who offers seeds of *Clarkia amoena* and its lilac-hued sidekick, *Clarkia pulchella* (Latin, pretty, literally 'beautiful and tiny'). Garden encyclopedists who state that "all of the hybrids are superior to the wild species" have let the thrill of genetic meddling overpower their ability to see simple beauty.

CLEOME

Common name: Spider flower

Genus: Cleome, derivation uncertain; the name was used by Theophrastos, an early Greek writer about plants, perhaps from *kleos* Greek, glory, fame.

Family: Capparaceae, the caper family < *capparis* Latin < *kapparis* Greek, the name of the prickly Mediterranean shrub, *Capparis spinosa*, whose young flower buds are picked and pickled to make the capers we sprinkle in various foods as a relish. The English put them in caper sauce to disguise the reek of boiled mutton; the French strew capers over a *salade niçoise*.

SPECIES

Cleome hassleriana (Botanical Latin, named after its discoverer and first collector Emile Hassler) native to tropical America is sometimes sold as *Cleome spinosa* (Latin, with spines, referring to the prickly stalk). The long, delicate stamens like spidery legs account for the popular name of this easy-to-grow annual that can reach two metres. A few rows of Cleome look as spiffy as plume-hatted soldiers standing sentinel at the back of a garden border.

∾

COCKSCOMB

Other names: Wool Flower, Red Fox (Brit.), Plume Plant, and—seen in one desperate greenhouse—Jolly Plumed Knight! King Arthur would have reconsidered his breakfast all over the Round Table, had any knight ever presented himself in so vulgar a motley of garish colours.

Genus: Celosia < *keleos* Greek, burning, flaming, in reference to the grotesque red inflorescence that bears a slight resemblance to the roseate combs of roosters, hence the most familiar common name

Family: Amaranthaceae < *amaranthus*, Latin < *amaranton* Greek< *a* not + *marantos* fading. Many species have flowers and coloured leaves that retain their colour for months.

SPECIES

This large genus of herbs from the Asian tropics supplies several bizarre but popular novelties to the trade, including *Celosia plumosa* (Latin, plumed) in which the untidy, chaffy flower heads glow in blatant, hideous oranges, stomach-churning metallic yellows, and supersaturated reds of halucious vulgarity. *Halucious* is an Anglo-Yiddish adjective, from a Hebrew verbal root that means 'to faint.' Something halucious is so bad of its kind that its ugliness or sheer, awful shlockiness makes you feel faint. Good word. Spread it, gardeners.

The common cockscomb of horticulture, *Celosia argentea cristata* (*cristata*, Latin, with a crest, referring to the flower head) is a variety of an Asian weed, *Celosia argentea* (Latin, silvery, referring to the flower spike) and grows with the revolting gusto of a weed. True to its amaranthine family, the damn flower spikes never fade, but sit there in the garden screaming colour, like tiresome guests in glow-in-the-dark Hawaiian shirts who will not turn down the volume of their portable CD player that blares gangsta rap hits like "Yo, I Jes' Offed A Granny in the Parking Lot; Shit, Found Out Later She the Welfare Snot!"

∾

COLEUS

Other common name: The Foliage Plant

New Genus: Solenostemon < *solen* Greek, pipe, tube + *stemon* Greek, stamen. The genus name refers to the stamens, the male reproductive organs that bear the pollen, which in this genus unite into a tube at the base of the corolla, the inner set of leaves that enclose and protect the stamens and pistils. While the minutiae of botanical name changes do not concern us here, this is a genus which scientific re-examination has shown to be suspect. Coleus was not distinct from Plectranthus as a genus, and the name (see below) was considered a misnomer, as the stamens are not truly enclosed by a sheath (see new name above).

Old Genus: Coleus, Botanical Latin < *koleos, koleon* Greek, sheath; an incorrect description of the way in which the stamens are enclosed

Family: Labiatae, the mint family < *labiatus* Botanical Latin, lipped, with a prominent lip < *labia* Latin, the lips. In botany, the labium is the lower lip of a flower with two lips. In flowers of the mint family, this labium is highly developed and enlarged, while the upper lip of the corolla though present is rudimentary. Present but rudimentary: that sounds like me in math class. The mint family is large, with 224 genera and 5,600 species in the tropics and warmer temperate zones.

SPECIES

Pattern mitigates gaudy colour, and coleus exemplifies this old garden saying. Okay, I wrote it last week. At home in Javanese jungles and islands of the South Pacific, coleus burgeons with brilliantly patterned leaves whose deep, rich hues make this a stand-out annual in any summer border. The parent plant of most commercial coleus is *Solenostemon scutellarioides*. The specific epithet means 'like another plant called *Scutellaria*,' also a member of the mint family, whose ripe calyx, the outer set of modified leaves that enclose the base of a flower, looks like a small dish (*scutella*, Latin, small dish). *-Oides*, the suffix of the specific, like *-oid*, a common ending on English adjectives, means 'formed like' and is derived from *eides*, a Greek word meaning 'form' or 'shape.' The buyer will still see these plants labelled *Coleus* in most nurseries.

❧

Common Name: Tickseed

COREOPSIS

Genus: Coreopsis < *koris* Greek, a bedbug, a tick + *opsis* Greek, vision, seeing, a looking at; and so used to suggest resemblance, thus coreopsis is an attempt to form a word that means 'bug-like' by some botanist acquainted with a list of Greek roots, but having no knowledge at all of how Greek forms words. 'Bug-like' in Greeked English would be better formed as *corioid*. The seed container of these plants looks like a little bug.

Family: Compositae, the daisy family

SPECIES
About a dozen annual species of coreopsis and their varietals produce cheerful yellow flowers, easy to grow from seed, which bloom all summer and make long-lasting cut flowers.

❧

Common names: Bachelor's Button, Bluebottle, Ragged Sailor

CORNFLOWER

Genus: Centaurea < *Kentauros* Greek, a mythical half-man, half-horse, a centaur who galloped into the Greek imagination from Thessaly where the name probably evolved from *kentron* Greek, spur. Thus the name originally meant 'he who spurs horses on.' When centaurs were feeling poorly, they were said to nibble on cornflowers and be suddenly restored to studly vigour. The genus also includes Dusty Miller and Basket Flower.

Family: Compositae

SPECIES

Centaurea cyanus (Botanical Latin < *kyanos* Greek, dark-blue, referring to the flowers). Cornflower has some claim to being the most widely grown annual in the world.

From Sicily came Dusty Miller, *Centaurea cineraria* (Botanical Latin, grey as ashes < *cinis, cineris* Latin, ashes, dead coals). Ash-grey, hairy down on the leaves makes this plant look like *Senecio cineraria*.

WORD LORE

Cornflowers grow wild in the grain fields of southern Europe, hence the common name repeated in many European languages. Bachelor's Button arose as a common name in Victorian times when one plucked cornflower made a cheap boutonniere for the buttonhole in a single gentleman's lapel. Basket Flower is so called because the flower buds of *Centaurea americana* look like they are inside little baskets.

The specific epithet of *Centaurea cineraria* has related words in English and French. *Cinis, cineris* (Latin, ashes) gives rise to: incinerator 'device to burn things to ashes,' *cendrier* French, ashtray, sinter 'iron dross' and cinders. But perhaps best known of all is the fairytale of the poor little girl who had to sweep the ashes from the fireplaces of the rich. In the French of Charles Perrault's classic children's story she was *Cendrillon* (little ash), in German *Aschenbrödel* (stirrer-up of ashes), and in English Cinderella, then later in a male version in a Hollywood film starring Jerry Lewis, *Cinderfella*.

∾

COSMOS

Genus: Cosmos < *kosmos* Greek, harmony, good order, hence the ordered harmony of the world and the universe, hence derivatives like cosmos, microcosm, cosmic, cosmopolitan. Cosmetic derives from an occupational name in ancient Greek, *kosmetes*, one who arranges, from *kosmein*, to put in order. *Kosmetes* often referred to a male slave who adorned a king or prince, one who arranged hair in fact, and this nuance leads to our present use of cosmetic and cosmetician.

Family: Compositae, the daisy family

SPECIES

The species which is the parent of most commercial varieties comes from Mexico, *Cosmos bipinnatus* (Botanical Latin, twice pinnate) where pinnate is a technical term in the vocabulary of leaf shapes. It means arranged like a feather (*pinna* Latin, feather) with leaves on each side of a common stalk. Bipinnate means these leaves are themselves divided and have their own leaflets. *Cosmos sulphureus* (Botanical Latin, of sulphur-yellow colour) varieties at one metre are shorter than the two-metre *bipinnatus*, and they flower earlier.

Other common names: Cathedral Bells, Mexican Ivy

Genus: Cobaea < named after a Jesuit priest, Padre Bernardo Cobo (1572-1659). He was a Spanish missionary in Mexico and Peru where he collected botanic specimens including samples of this splendid tropical vine.

Family: Polemoniaceae, the Jacob's Ladder family, named after its typical genus, polemonium, a perennial herb with deep-blue flowers. The plant was named after one of medicinal use mentioned in the writings of the scholar Polemon of Cappadocia, who flourished around 180 B.C. as a writer of travel guides and antiquarian studies in which he delighted readers by constantly correcting his predecessors.

SPECIES

Cobaea scandens (Latin, climbing) clambers up rough surfaces by means of airy tendrils at the ends of its leaves. The purplish, bell-shaped blossoms look vaguely like a cup on a saucer.

The Latin verb *scandere* 'to climb' whose present participle is used adjectivally in the specific name of this plant has derivatives and related words in modern English. To scan a line of poetry was first to tap out the rhythm of the poetry with one's feet on the floor, that is, by a 'climbing' motion as one's foot was raised up and down. The sense of scan meaning 'to look at, to examine' develops from poetic scansion. In Greek a *skandalon* was something put in one's way that one had to climb over, that is, a stumbling block. In the middle ages,

CUP-AND-SAUCER VINE

Roman Catholic theologians borrowed the Greek word into church or ecclesiastical Latin as *scandalum* and applied it to a moral stumbling block, anything that caused one to stumble into sin. Later, in the sixteenth century, scandal meant any act that brought religion into discredit, such as a celibate bishop fathering twelve children. Such scandalous behaviour gives a hint of the modern use of the word. Through Old French a variant of the Latin *scandalum* came to mean the uttering of false charges of sin. That variant became the word *slander* in modern English.

∾

DAHLBERG DAISY

Other common name: Golden Fleece

Genus: Thymophylla < *thymon* Greek, thyme, the familar aromatic herb of cooking, but also used by the ancient Greeks to add a resinous perfume to burnt sacrifices, hence its root in the verb *thyein* 'to burn as a sacrifice to the gods'+ *phyllon* Greek, leaf. This twenty-centimeter-high daisy of Texas and Mexico has finely divided leaves that give off an aroma similar to thyme.

Family: Compositae, the daisy family

SPECIES

Thymophylla tenuiloba (*tenuis* Latin, thin + *lobus* Latin, lobe, referring to its slender seed pods) gets its common name from one of the Dahlberg brothers, Nils or Carl Gustav, who in the eighteenth century were Swedish friends of Linnaeus, the great regularizer of plant names. This is a bright yellow daisy for the sandy loam of a hot, sunny garden.

∾

DAHLIA

Genus: Dahlia < Doctor Anders Dahl (1736-1820), a pupil of Linnaeus, gets immortalized here. After the death of Linnaeus, Dahl tried to have the great botanical collections amassed by his teacher kept in Sweden, but they were sold to English buyers.

Family: Compositae, the daisy family

SPECIES

Most Dahlias are tuberous-rooted perennials from the highlands of Guatemala and Mexico. But early-flowering dwarf Dahlias will bloom from seed the first year, and are treated as tender annuals.

∾

Common names of several species: Jimsonweed, Thorn-Apple, Angel's Trumpet, Devil's Trumpet, Stramonium, Metel

The leaves, seeds, and most plant parts of all Datura species are toxic, narcotic, stupefying, hallucinogenic, and sometimes fatally poisonous to humans, pets, and livestock. *Angel's Trumpet* indeed! Datura is listed in *Poisonous Plants of Canada* (Agriculture Canada, 1990) where the authors give the published evidence of sickness, poisoning, and death resulting from "ingestion of plant parts." This genus has no place in my garden, yet I see it in two-metre-high clumps near children's playgrounds here in southern Ontario. And how would you like Fluffie, your beloved Persian cat, rushing about the livingroom on a bad Datura trip, being pursued by giant, kaleidoscope-eyed purple mice with rotary claws?

Genus: Datura as a word was borrowed into English from one of the Sanskrit-based languages of India. *Datura* or *dewtry* Early Modern English < *dhatura* Hindi & *dhutra*, *dhotra* Marathi < *dhattura* Sanskrit, Thorn-Apple. About twenty species occur so that the genus is distributed worldwide.

Family: Solanaceae, the nightshade family

WORD LORE OF VARIOUS SPECIES

Jimsonweed

Datura stramonium is Jimsonweed, *stramoine commune*, a pest many Canadians think belongs only to dime novels and old movies about the American West containing dialogue such as this sample I made up, after a wayward youth frittered away on too many Saturday afternoons watching Gene Autry, Roy

DATURA

CAUTION:

Rogers, and wonderful character actors like George "Gabby" Hayes speak lines such as these: "Zeke went plumb loco, sheriff. Chewed him a big wad of jimson, got all bunged up, and tried to give hisself one of them there enemies with that cactus from the front window of the saloon. Next morning Zeke wakes up and there's his asshole down around his ankles. Course now, Zeke's always bin pretty darn quick on the uptake—right smart feller most times. Yessir, he took a gander at them innards of hisn, and Zeke jes' knowed he were in trouble."

In fact, Jimsonweed now grows wild in waste places and along roadsides through southern Canada from Prince Edward Island right into British Columbia. Jimsonweed is a dialect variant of Jamestown Weed, and is one of the few common names for plants that we can trace back to one specific incident, so startling to early settlers that it was talked about for decades, and thus it gave rise to the plant's common name.

Jamestown in what became the state of Virginia was the first permanent English settlement in America, begun in 1607 as a trading post on the James River named after King James I. It was there that Captain John Smith made friends with an aboriginal woman named Pocahontas and her father Powhatan. Ships sailing to "London's Plantation in the Southern Part of Virginia" to repopulate and reprovision the colony were wrecked off Bermuda in 1609, and Shakespeare used an account of that disaster in writing *The Tempest*. Virginia tobacco as an export saved the colony, becoming the most popular smoke in England by 1618. But it was an incident later in the colony's history that gave rise to the name Jamestown Weed. This was Nat Bacon's Rebellion in the late 1670s. Nathaniel Bacon, a plantation owner, led a group of militia against friendly Indians after one of his foremen was murdered in the field. Bacon then went on to try to capture the whole colony and redress various wrongs. For a few months in fact, his forces held most of the Virginia colony. But the governor called in British Red Coats and Bacon was defeated, dying of dysentery before a battle. After his death Nat Bacon became a kind of colonial folk hero, and the following Jamestown Weed incident was widely told. The first report of it in print was in 1705 in Robert Beverly's *History of the Present State of Virginia*. *State* in the title means "condition."

The *James-Town* Weed (which resembles Thorny Apple of Peru, and I take to be the Plant so called) is supposed to be one of the greatest Coolers in the World. This being an early Plant, was gathered very young for a boiled Salad, by some of the Soldiers sent thither, to pacifie the Troubles of *Bacon*; and some of them eat plentifully of it, the Effect of which was a very pleasant Comedy; for they turned natural Fools upon it for several Days: One would blow up a Feather in the Air; another would dart Straws at it with much Fury; and another stark naked was sitting up in a Corner, like a Monkey, grinning and making Mows [moues] at them; a Fourth would fondly kiss, and paw their Companions, and snear in their Faces, with a Countenance more antick, than any in a Dutch Droll. In this frantick Condition they were confined, lest they should in their Folly destroy themselves … A Thousand such simple Tricks they played, and after Eleven Days, returned to themselves again, not remembering any thing that had passed.

Datura

In a well-known hippie-dippie drug text of the late 1960s, *The Teachings of Don Juan: A Yaqui Way of Knowledge*, author Carlos Castaneda reported rubbing himself with and drinking various extracts of *Datura inoxia*, which induced wild hallucinations. Angel's Trumpet, *stramoine parfumée* in French, has a specific *inoxia* that means 'not spiny.' There has been some reclassification and renaming in the Datura genus. *Datura inoxia* is now *Datura meteloides*.

Metel
Datura meteloides contains one of the old names for a Middle Eastern species of Datura. It entered English in the sixteenth century as metel < *methel* modern Latin < *jaus matil* Arabic, literally 'incomparable nut' referring to the thorny fruit of *Datura stramonium*. It was matchless and incomparable because of the hallucinations it produced—if one were lucky enough not to be poisoned while trying to get high.

Stramonium
This word for a Datura species of Europe and Asia first appears in 1542 in a herbarium written by Leonhart Fuchs (1501-1566), the German physician and herbalist after whom Fuchsia is named. He says the word is Italian, probably because it looks Latinate. *Stramonium* is a Latinized variant of *durman*, an early Russian word for Datura, borrowed into Slavic languages from the Tartar language of Kazanistan where its form was *turman*. Perhaps noteworthy is the fact that the oldest plant names we have are for those plants that provide food and dope!

Stramonium was influenced in its form by true Latin words like *stramen* and *stramentum* 'litter, straw,' which the roots of some Datura species were said to resemble.

Thorn-Apple
This English common name refers to the prickly fruit of Datura species.

Brugmansia
Some of the tree and shrub forms of Datura have been reclassified and given their own genus.

Brugmansia arborea is a tree form once called Peruvian Thorn-Apple. *Brugmansia sanguinea* of South America has a blood-red corolla. Sebald Justin Brugmans (1763-1819) taught botany and zoology at the University of Leiden.

Historical & Modern Uses
Datura was said to be part of the prophetic potion consumed by priests at the most famous oracle in ancient Greece at Delphi. In

India, *Datura meteloides* was used in Ayurvedic herbal remedies against certain forms of insanity, but one has to question if such a cure was worse than the uncured malady.

∾

Common names of various species: Pink, China Pink, Clove Pink, Annual Carnation, Sweet William, Stinky Billy, Gillyflower, Clove Gillyflower, Sops-in-Wine

Genus: Dianthus < *Zeus*, chief god of the ancient Greeks, whose name had the irregular genitive form *Dios*, shortened to *Di-* + *anthos* Greek, flower, hence flower of Zeus. There are more than three hundred annual and perennial species and varieties in this genus.

Family: Caryophyllaceae, the pink family, named after its typical species, *Dianthus caryophyllus*, clove pink. *Karya* Greek, walnut + *phyllon* Greek leaf + *aceae* familial suffix of botany. Walnut leaves have a sharp aroma; so do the leaves of clove pinks. The family has at least 75 genera and 1,200 species.

French: *oeillet* 'little eye'

DIANTHUS

WORD LORE OF COMMON NAMES

Carnation < *caro, carnis* Latin, flesh, meat. *Carnatio* is a post-classical Latin word that first meant 'fleshiness, corpulence' (420 A.D.) In Middle English, Christian writers used it in 1410 as a contraction of incarnation. Then in the technical vocabulary of Renaissance Italian painting, *carnatio* or its Italian equivalent *carnagione*, came to mean the colour of human flesh as painted, either a light rosy pink or a deeper crimson tone. This applied easily to flowers that were red and pink like the Dianthus species called Carnations in English by 1578. But the art term continued to be used too. Here is John Harris writing in *Lexicon Technicum, or An Universal English Dictionary of Arts and Sciences* (1704-10): "*Carnation* is a term in Painting, signifying such Parts of an Human Body as are drawn naked ... and when this is done Natural, Bold, and Strong, and is well coloured, they say of the Painter, that his Carnation is very good."

Although the flower came in white and many reds less than bright, the flower name stuck, and poets loved the word. Byron speaks of a thing "carnation'd like a sleeping infant's cheek" and his hero Don "Juan grew carnation with vexation."

But one intriguing etymology, labelled with all due modesty "perhaps" in the *New Shorter Oxford English Dictionary*, claims that carnation came to be applied to the flower only through a mangling and misreading by an English herbalist of a word in an Arabic book about plants where clove pink or clove was *karanful*, itself an Arabic attempt to render the Greek name *karyophyllon*, now the name of the family in which carnations reside. This seems an outré entrée indeed for the pretty little carnation as an English flower name, when the Latin is so clear and evocative and supported by textual proof. However, even etymologists like to complicate matters.

Sometimes Occam's Razor must be applied in linguistics. It is the principle of cutting away excessive constituents in a subject being analysed, first enunciated—though not in these precise Latin words—by William of Ockham, the medieval scholar and philosopher. Occam's Razor is: *entia non sunt multiplicanda praeter necessitatem*. Facts or entities should not be multiplied beyond what is necessary.

The same Latin root found in carnation appears in words as diverse as carnival, carnivore, carrion, charnel house, crone, and reincarnation.

Carnations usually have as one ancestor the species *Dianthus caryophyllus*.

Pink

Pink has a curious history. The first printed evidence of the word in English is from 1503 where 'to pink' is a verb meaning to pierce, to thrust. As a noun referring to the flower, pink appears first in 1573, perhaps from the jagged margins of the flower petals on most species. Pinking shears which cut a saw-toothed edge in fabric carry the old verbal meaning. Not until 1678 is there in print a use of pink as a colour adjective meaning a pale, reddish hue. But even by Elizabethan times the flowers were popular in British gardens. In *Romeo and Juliet* (1592), Shakespeare has a character say, "Nay, I am the very pinck of curtesie." This must be a metaphorical use of the flower name, much as the word *flower* itself was used in phrases like "the flower of our bold soldiers" to describe the finest example of something. This very old use disappears from print for several centuries and then pops up again in the middle of the nineteenth century in a Cockney slang phrase indicating good health: "in the pink, guv!"

The colour pink became the traditional colour for the clothes of female infants, while baby boys wore blue. This may be why the Nazis made homosexuals waiting to be gassed in concentration camps wear pink triangles. Pink meant 'radical left' long before the publication of *The Communist Manifesto* in 1848. Commies were "Reds" in the United States, but calling them "Pinkos" made them sound not just communist but also effete.

Pinks derive from the species *Dianthus plumarius* (Botanical Latin, with a plume, referring to the fringed petals). Clove-scented pinks with the bitter white heel of the petal removed are still added to salads, soups, sauces, fruit dishes, vinegar. The spicy-sweet scent perfumes soap and is tinctured as an aromatic in perfume manufacture.

Sweet William

Sweet William & Stinky Billy

Sweet William was so dubbed in English by the mid-1500s, no doubt for some lost Bill of cheerful temperament. Some Scots did not like the flower, and Stinky Billy originated as a common name in the Scottish Highlands.

Sweet Williams descend principally from the species *Dianthus barbatus* (Latin, bearded, referring to the 'bearded' petals).

Gillyflower

Here is an example of English folk etymology, where English ears hear an incomprehensible word and try to make it understandable by turning the foreign sound into an explicable

English word. The old Greek name *caryophyllon* (still the family name of Dianthus species) went into medieval Latin as *caryophyllum*, then into Old French as *gilofre* and *girofle*, thence into Middle English as *gilliver* and finally *gillyflower*. Gillyflower sounded about right for a spicy little flower, since gillie meant a servant lad in Scotland and the north of England. Gillyflower which appeared first in English may even have influenced the creation of the later name, Sweet William. The spice English calls clove began life as Old French *clou de girofle* 'nail of gillyflower.'

Sops-in-Wine
Chaucer used this name for pinks, introduced into Britain by Norman monks around 1100 A.D. The clove fragrance of the flowers caused their frequent use in cookery and they were dried and sprinkled in wine to pep up its flavour, hence Sops-in-Wine.

DIGITALIS

Other common names: Annual Foxglove, Ladies' Fingers, with which compare the French common names below.

Genus: Digitalis (Botanical Latin, finger-like, referring to the tubular flowers) < *digitus* Latin, a finger, a toe. The celebrated German botanist Leonhart Fuchs (1501-1566) named the plant *Digitalis* by translating into Latin the first part of the then common and still common German name for the plant. *Fingerhut* means literally 'finger hat' hence 'thimble.'

Family: Scrophulariaceae, the figwort family, named after its typical genus, *Scrophularia auriculata* (Botanical Latin, with little ears, referring to the heart-shaped base of the leaves), which in turn was named because herbalists thought it would cure scrofula, a now rare, tuberculous inflammation of lymph nodes and glands once seen in children and adolescents. *Scrofula* is the Latin word for 'breeding sow.' The disease was so named because the inflamed glands in victims' necks made them look somewhat pig-necked. A synonym for scrofula was "The King's Evil" because many Renaissance Europeans believed the touch of a king could cure the disease. This was no idle belief. In his *History of England*, Macaulay reports that this regal practice attained hysterical frequency under England's Charles II who in his reign touched almost one hundred thousand of his afflicted subjects. The last British monarch to touch scrofula victims was Queen Anne. Dr. Samuel Johnson, the great writer and dictionary-maker, was touched for scrofula by Queen Anne. Johnson grumpily reported that Her Highness's majestical pinkies produced not the slightest alleviation of his symptoms.

Also in this scrophulariacious family are snapdragon, verbena, slipperwort, and monkey flower.

French: *digitale*, and French common names like *gant de Notre-Dame* 'Our Lady's glove' and *doigt de la Vierge* 'Virgin Mary's finger'

SPECIES

The traditional foxglove is a biennial, leafing the first year, flowering the second. But hybridizers have produced annual varieties of *Digitalis purpurea* (Latin, purple, referring to the flower colour), one of them called "Foxy" in the trade.

WORD LORE

Digitalis or foxglove

From Digitalis to Digital Info

The Latin word for finger, *digitus*, harks back to an Indo-European root **deik*, **deig* and its shorter forms **dik* and **dig*, all of whose basic meanings involve pointing to something; then meanings of the simple root expand, as is their wont through time and use, and are used to make words of showing, telling, and saying. So *digitus* is the finger as pointer. The distantly related Latin verb of saying, *dicere*, gives hundreds of English words like diction, dictum, dictionary, benediction, condition, judicial, prediction, vindicate, etc. In mathematics, the ten arabic numerals from zero to nine are digits, obviously from using the ten fingers to count. The plant name *Digitalis* entered English at about the same time in the early seventeenth century as did the adjective *digital* with its anatomical meaning 'pertaining to the fingers or toes' and its mathematic sense of 'pertaining to, using, or being a digit, a number.' In the twentieth century, digital expanded into new meanings in computerese such as 'operating by means of data expressed in digits.' The verb *digitize* arose, meaning 'to turn sonic and graphic information into digits and then to store these numerical codes.' There follow uses like digital audio, digital radio, and digital video.

Foxglove

Foxglove is an apt folk name and an old flower name appearing in Anglo-Saxon or Old English as *foxesglofa* by 1000 A.D. The glove reference is obvious as the flowers look like the finger-stalls of a glove. But why is this plant associated with foxes? The reason is lost, but it probably predates the arrival of Germanic tribes in Britain. Vikings speaking Old Norse or Old Scandinavian made a similar plant/animal connection. Compare the modern Norwegian word for digitalis, *revbjelle* 'fox bell.' The English poet John Keats, while sonneteering, observed, "Where the deer's swift leap / Startles the wild bee from the foxglove bell."

Medical Use of Digitalis Species

Certain chemicals called glycosides are extracted from the leaves and seeds of *Digitalis purpurea* and *Digitalis lanata* (Latin, woolly, referring to the racemes, the elongated clusters of white flowers in this species native to the Balkans). These include digitalis, digoxin, digitoxin, lantanoside C, acetyldigitoxin, and deslanoside. An older name was digitalin. They are cardiotonic agents that increase the force of heart muscle contraction without increasing oxygen use and are thus of great effect in many instances of congestive heart failure. Heart failure caused by high blood pressure and/or by hardening of the arteries due to deposits of fatty plaque in the larger arteries can be alleviated often by these chemicals which act in general to make the heart a more efficient pump. The first important medical paper on the value of digitalis was published in 1785 by William Withering of Birmingham, under the title, "Account of the Fox-glove and some of its medical uses."

Other common name: Flowering flax

FLAX

Genus: Linum < Latin, flax. This Mediterranean word of ancient lineage has relatives in most of the major Indo-European languages. Cognate with *linum* and thus not derived from it are Old English *lin*, Russian *lin*, Old Welsh *llin*, Greek *linon*, to name but a few of the reflexes of the Indo-European root **linom*, certainly one of the oldest fabric words in the West. The ancient Egyptians wove linen cloth and called it *meni*. The historical uses of linen give rise to many current English words like line, originally a thread of flax fibres, then cord, string, fishing line, and finally a drawn line. Lint was first wisps of linen imperfectly woven and hence detachable. A small bird fond of flax seeds became known as the linnet. The product name *linoleum* combines *linum* 'flax' + *oleum* 'oil,' being first made partially from solidified linseed oil.

Family: Linaceae, the flax family, native to the shores of the Mediterranean Sea

SPECIES

Linum usitatissum (Latin superlative adjective, most useful) or common flax is the source of linseed oil and of linen. With modern improvements and machines, linen cloth is still made much as the ancient Egyptians made it, from the stem fibres of annual flax. In Canada flax is grown as an oilseed crop in Manitoba, Saskatchewan, and Alberta. Linseed oil is used in Canada's domestic paint industry. The seedmeal remaining after oil is extracted makes a proteinaceous animal feed. Flax fibres are part of Canadian paper money where the linen threads add strength and toughness to the paper. Flax straw makes cigarette paper. The linen species also makes a blue-flowered garden annual.

   Linum grandiflorum (Latin, big-flowered) is the best garden subject, and can be sown every few weeks into summer, to produce a continuous show of this plant whose flowers are perky only for a day or two.

WORD LORE OF FLAX

Old English *fleax* is cognate with modern German *Flachs*, both related to *flechten* 'to weave, to intertwine,' ultimately going back to the Indo-European root **plek* which, along another etymological sidepath, gives words like pleach.

❧

FLOWERING TOBACCO

Genus: Nicotiana < named after Jean Nicot (1530-1600). Nicot was the French ambassador to Lisbon in 1560 when he received tobacco seeds from Portuguese sailors returning from the American tropics. That same year he sent a sample of leaves to Catherine de Medici, then Queen of France who introduced tobacco to the French court in the sixteenth century. The result centuries later was a worldwide addiction to cancer-causing tobacco-smoking.

Nicotiana

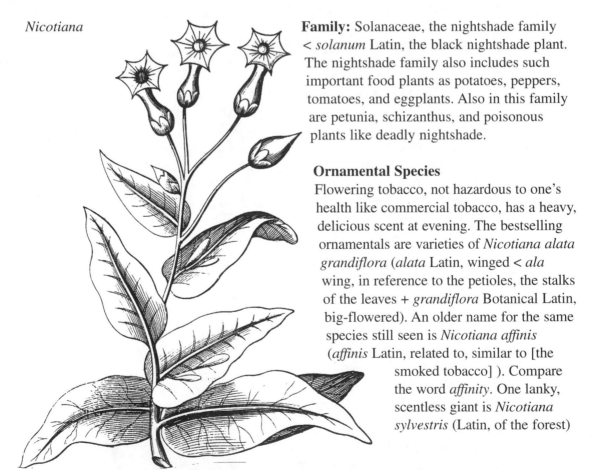

Family: Solanaceae, the nightshade family < *solanum* Latin, the black nightshade plant. The nightshade family also includes such important food plants as potatoes, peppers, tomatoes, and eggplants. Also in this family are petunia, schizanthus, and poisonous plants like deadly nightshade.

Ornamental Species

Flowering tobacco, not hazardous to one's health like commercial tobacco, has a heavy, delicious scent at evening. The bestselling ornamentals are varieties of *Nicotiana alata grandiflora* (*alata* Latin, winged < *ala* wing, in reference to the petioles, the stalks of the leaves + *grandiflora* Botanical Latin, big-flowered). An older name for the same species still seen is *Nicotiana affinis* (*affinis* Latin, related to, similar to [the smoked tobacco]). Compare the word *affinity*. One lanky, scentless giant is *Nicotiana sylvestris* (Latin, of the forest) whose pale white flowers make a nice spray at the back of the border. Aphids love the sticky leaves and stems of Nicotiana, perhaps for the same reason that nicotine-stained human wretches love it.

WORD LORE OF TOBACCO

Nicotiana tabacum is the source of the dried tobacco leaves that are smoked. Taino is an extinct language of an aboriginal Arawakan people of the Lesser Antilles and the Bahamas who wrapped tobacco leaves in a loose, cigar-like roll and also made a pipe to smoke it. They used a word that Spanish sailors heard as *tabaco* to name the cigar and the pipe. Farther south in the Amazon basin are the Tupi people who have *taboca* to mean any reed-like plant. The Taino may have named themselves because their people grew tobacco. On the island of Haiti, tobacco was called *taino*. The form of the Spanish and thus of the English word may have been modified and influenced by the name of the island of Tobago and by *tabbaq* 'medicinal herb,' an unrelated Arabic word perhaps learned from the Moors and known by some of the Spanish discoverers.

∽

Genus: Myosotis < *myosotis* Greek, mouse-ear < *mys, myos* Greek, mouse + *ous, otos* Greek, ear. Myosotis was an ancient name for a plant of an entirely different genus whose little, pointy leaves reminded someone of the ears of mice. Even the ancient Greeks could get too damn cute.

Family: Boraginaceae, the borage family, a small group of European and Asian herbs. Borage < *bourrache* Old French < *borrago* Medieval Latin < *abu 'araq* Arabic, father of sweat, because early Arabic medicine used borage extract to induce sweating

SPECIES

Myosotis sylvatica (Botanical Latin, of the woodland < *silva* Latin, forest) and *Myosotis alpestris* (Botanical Latin, originally of the lower Alps, now: of any lower mountainous regions) are two species sold as seed. More than seven different plants are called Forget-me-not in English. Myosotis with its blue, pink,

FORGET-ME-NOT

or white flowers is a cheerful self-seeder and has varieties that are annual, biennial, and perennial.

WORD LORE OF GREEK AND LATIN *MICE* WORDS

The Greek word for mouse, *mys*, that helps give the botanical name of the Forget-me-not is related to the Latin word for mouse, *mus*. This Latin word formed the basis for the term *muscle. Musculus* Latin, a little mouse, then a muscle < *mus* mouse + *-culus* diminutive noun suffix, little, small. The Romans thought muscles rippling under the skin looked like *musculi,* little mice. This playful image probably occurred while watching workers flexing certain muscles, such as the biceps of the arm. The same metaphor occurred to ancient Greeks, from whom the Romans may have borrowed the metaphor. Greek *mys, myos* meant mouse, then muscle, and appears in numerous medical terms, two of which are:

Myocardium

mys, myos muscle + *kardia* Greek, heart
Myocardium is heart muscle, cardiac muscle that surrounds the heart in a tough, thick layer.

A myocardial infarction is a heart attack due to closing off of a coronary artery that causes an infarct of heart muscle. Part of the muscle dies from lack of oxygen.

Myometrium

mys, myos Greek, muscle + *metrium* Latin form of *metra* Greek, uterus

The myometrium is the smooth-muscle layer of the uterus that surrounds the endometrium. It contracts the uterus during menstrual cycles and during birth.

Muscle has a variation in Old English *musle* and then Middle English *musselle* to name a marine mollusk, the mussel, whose two valves are closed by a strong muscle.

Latin *mus* and Greek *mys* both derive from the Indo-European root **mus-* which appears also in a related language of ancient India called Sanskrit where mouse is *mus*. The Sanskrit *mus* has a derivative *muska*, literally 'the mouse place of the human body, the crotch where pubic hair has a mouse-like appearance,' then the meaning expands to refer directly to the scrotum or the vulva. Old Persian borrowed that Sanskrit word to name an extract from the sac under the abdomen of certain male deer, and Persian *mushk* became eventually musk in English to name a perfume from this dried extract.

∾

FOUR-O'CLOCKS

Other common name: Marvel of Peru

Genus: Mirabilis < *mirabilis* Latin, wondrous, marvelous, miraculous < *mirari* Latin, to be amazed at + *abilis* Latin adjectival suffix, able, capable of; hence the many English words from these roots, like admirable and miracle (*miraculum* Latin, literally 'a little something to be amazed at')

Family: Nyctaginaceae, the four-o'clock family < *nyx*, *nyktos* Greek, night + *aginemenai* Greek, to bear fruit, to bloom; hence a family of night-bloomers. This tropical group of herbs, shrubs, and trees has only three species used for ornamental purposes: the four-o'clocks, the sand verbenas, and those splendid vines, the Bougainvilleas.

SPECIES
Mirabilis jalapa (Botanical Latin, of Xalapa in Mexico, where Spanish plant collectors first saw them in the sixteenth century and shipped seeds back to Madrid introducing the plant to European gardens) is the familiar species used as a garden annual. Its flowers open in late afternoon and stay

open all night, closing before noon the next day, to facilitate pollination by moths during a tropical night. In North America, hummingbirds lovingly tongue the tubular blossoms. The flowers are yellow through various reds and send forth sweet scent at twilight. Victorian gardeners in southern England made this tender tropical immensely popular a century ago. Four-o'clocks are easy to bloom from seed in one season, or dig up the fat tuberous roots in the fall and store them over winter to obtain a very large shrub indeed during its second summer.

FUCHSIA

Fuchsia

Old common name: Lady's Eardrops

Genus: Fuchsia < named to honour Leonhart Fuchs (1501-1566), German physician, herbalist, writer of several popular sixteenth century books about plants, herbals adorned with exquisite woodcuts. Fuchs has no historical connection with the genus. His surname is the German word for fox, chosen either to nickname a red-haired ancestor or to refer to the identifying sign on an ancestor's house. In Medieval Europe, long before literacy, numeracy, and street numbers, houses bore animal signs, so a German surname might begin as *Hans zum Fuchs* 'Hans at (the sign of) the Fox.'

Family: Onagraceae, the Evening Primrose family < Onagra, former botanical name of the family < *onagra* Greek, a female wild ass < *onagros* Greek, male wild ass = *onos* Greek, donkey, ass + *agrios* Greek, wild; but why the name? Because some wandering botanist of yore saw wild asses eating evening primroses? We shall never know. The most important genus of Onagraceae in commercial horticulture is Fuchsia. Other genera include garden subjects like Clarkia.

SPECIES

How do we dare to include a 100-species genus of perennial shrubs in this part of the book? Because most Canadian gardeners use Fuchsia as an annual, bought in a plastic basket, hung up to beautify porch or patio and to attract hummingbirds, then slain by frost.

Fuchsia procumbens (*pro* Latin, in front of + *cumbens* < a participial form of *cubere* with an infixed *m*, Latin, reclining, lying down; hence *procumbens* means prostrate, growing flat across the ground) is also called trailing Fuchsia, the parent of most hanging-basket Fuchsias.

Fuchsia magellanica with red and blue flowers is native to southern Chile and southern Argentina in the vicinity of the Straits of Magellan, the body of water named after Ferdinand Magellan, the great Portuguese navigator who sailed through the straits in 1520 as part of the first circumnavigation of the world. Magellan's Fuchsia is one of the parents of most commercial Fuchsia hybrids.

∽

GAILLARDIA

Other common names: Blanket Flower, Indian Blanket (because its colours reminded gardeners in the American southwest of the bright hues of Navajo blankets)

Genus: Gaillardia < to honour Gaillard de Charentonneau, eighteenth century French jurist and supporter of botanical research. The genus is native to southwestern North America.

Family: Compositae < *compositus* Latin, placed together. The largest family of flowering plants is named after the compound flowers of its members. Small florets of individual flowers make up large clusters or heads. The standard example of Compositae is the daisy.

SPECIES

Gaillardia pulchella (Latin, pretty, literally 'beautiful and tiny') is the garden annual whose seedheads can be dried for winter arrangements. *Gaillardia aristata* (Botanical Latin, bearded < *arista* Latin, ear of corn, ear of any grain) is a perennial species. 'Bearded' is a technical reference to the awns, a little crown of hairs below the stigmas of the flower. An awn is a hair-like bristle usually seen on the spikelets of grasses like oats.

∽

GAZANIA

Common name: Treasure Flower; the perennial species are sometimes called African Daisies.

Genus: Gazania < to honour Theodoros of Gaza (1398-1478), medieval translator of the botanical works of Theophrastos from Greek into Latin. Gaza was a small but celebrated city of ancient Palestine. This genus of about two dozen species is native to South Africa.

Family: Compositae, the daisy family < *compositus* Latin, placed together. The largest family of flowering plants is named after the compound flowers of its members. Small florets of individual flowers make up large clusters or heads.

SPECIES

Gazania longiscapa (Botanical Latin, with a long scape) is the annual. A scape is a leafless flower stalk or stem that arises at the ground, from *scapus* Latin, stem, stalk, shaft, trunk. Compare its diminutive, used in Roman anatomy, *scapula* 'little shaft,' then shoulderblade.

HELIOTROPE

Common names: Cherry Pie, Turnsole (a translation of its Greek name)

Genus: Heliotropium < *heliotropion* Greek, a species of heliotrope < *helios* Greek, sun + *trope* Greek, a turning. Contrary to medieval belief, the flowerheads do not turn toward the sun or follow the sun across the sky all day. Most annual herbs however do display some phototropic response (*phos*, *photos* Greek, light + *trope* turning), positioning their leaves for maximum photosynthesis.

Family: Boraginaceae, the borage family, a small group of European and Asian herbs. Borage < *bourrache* Old French < *borrago* Medieval Latin < *abu 'araq* Arabic, father of sweat, because early Arabic medicine used borage extract to induce sweating

SPECIES

Heliotropium arborescens (Botanical Latin, lit. 'tending to become tree-like,' hence woody or shrubby < *arbor* Latin, tree) is a perennial herb that may reach a height of 2.5 metres in its native Peru. In northern gardens, heliotrope is grown as an annual for its deep purple flowers and their rich aroma of fresh vanilla. The old English name, Cherry Pie, is a farfetched allusion to the dense, mature blossom.

GREEK MYTH

Heliotropion is the name of a Greek plant, mentioned as early as the writings of Theophrastos (371-287 B.C.), the philosopher and botanist who headed the first Academy after its founder Aristotle died. The plant's name was explained as "turning to the sun" and a mythological transformation story added. Apollo, the sun god, was an arrogant divinity, even for a Greek god. And, let's face it, the male Greek gods were as notable a crew of macho swaggerers as ever lolled on an Olympian cloud bragging of their sexual conquests. When not shining radiantly like the sun, Apollo seems to have spent many an eon beating adoring females off with a stick. Perhaps they got in the way of his diddling of shepherd boys? For whatever reason, Apollo burned many a maiden. One such was a smitten nymph, Clytia, whose amorous swoonings Apollo rudely ignored. Poor Clytia took to following the god around, mooning bovinely, and often just plunking herself down in a meadow and watching Apollo as he raced across the heavens each day in his blazing yellow chariot which was the sun. Clytia apparently expired in this stance— perhaps from meadow damp or terminal neck crick—and some of the other gods took pity on her. Ever notice how those old gods always waited until one was dead before the pity flowed? Some things don't change. In any case, Clytia was transformed into a heliotrope, so she could follow the progress of Apollo all day forever. Like, get an afterlife, girl.

IMPATIENS

Other common names of various species: Balsam Flower, Balsam Impatiens, Busy Lizzie, Impatience, Jewelweed, Patient Lucy, Snapweed, Touch-me-not or its Latin translation, *Noli-me-tangere.* Himalayan Balsam is called Policeman's Helmet in Britain.

Genus: Impatiens < Latin, impatient < *in (im)* Latin, not + *pati* to put up with, to endure. The genus was named because the ripe seed pods split open explosively and discharge seeds when touched, hence some of the common names like Touch-me-not, Busy Lizzie, and Snapweed. This dispersal ploy has worked well for Impatiens, which is a large genus of more than 500 species widely distributed in Asia, tropical Africa, and North America.

Family: Balsaminaceae, the balsam family. Balsam and balm are related, evocative words that have travelled far, beginning perhaps in ancient Egypt in a word like *m'aam* 'embalming spice,' one of the spices used to mummify corpses. Note that the root still resides in our English verb *embalm*. Egyptian *m'aam* or one of its related forms like *m*(b)*'aam* was then borrowed by or is cognate with Hebrew *basham* and Arabic *balasan* 'balsam.' The Greeks borrowed the Arabic as *balsamon*, from which the Romans made *balsamum*. The Latin form evolved into Old French *basme*, giving Middle English *baume* and modern French *baume* and modern English *balm* and *balmy*. A balmy wind was originally a healthy wind. To preserve a dead body "in balm" and other spices and by additional means was, in Old French, *embasmer*, later *embaumer*, which entered English as embalm. "Not all the water in the rough, rude sea / Can wash the balm from an anointed king," crowed Shakespeare's Richard II, illustrating the Elizabethan use of balm to mean any fragrant oil or perfume. Balsam too meant any resinous ointment that preserved, healed, or soothed. For more details of balsam's etymology, see under *Fir* in "Trees Native to Canada."

GARDEN SPECIES

Because impatiens blooms all summer long and readily in shade, some hybrids have become the bestselling bedding plants in North America.

Impatiens balsamina (Botanical Latin, producing balsam) has a specific name that recalls uses of the juice of various species rubbed on the skin for reputedly curative purposes. For example, some of the sting of poison ivy or stinging nettle can be alleviated by rubbing impatiens juice on the inflammed area. Garden balsams are easy to grow, and do encourage young children fresh to gardening to touch the ripe pods in the fall and enjoy their popping spray of seeds. This much-hybridized species is native to southern India and China.

Impatiens wallerana is grandma's Busy Lizzie, named to honour the Reverend Horace Waller (1833-1896), a missionary and plant collector in central Africa.

Impatiens hawkeri is New Guinea Impatiens, collected on South Sea voyages by a Lieutenant Hawker.

NATIVE SPECIES

Impatiens biflora (Botanical Latin, twin-flowered) is our tall Jewelweed. Snapweed, also of North America, is *Impatiens pallida* (Latin, pale-colored, said of the flower).

Other common names: Shrub Verbena, Yellow Sage

LANTANA

Genus: Lantana < a Late Latin name for a species of Viburnum transferred to this aggressively weedy tropical shrub because their flowerheads are vaguely similar < *lantana* Italian dialect, the wayfaring tree, *Viburnum lantana*, probably < *lanthano* Greek, I hide, I escape notice. The dense habit of growth of the wayfaring tree allowed travellers to hide from thieves under its cover (?). The Italian *lantana* may also be a contracted then extended folk form of *lanatus* Latin, woolly, referring to the bushy undergrowth of this viburnum.

Family: Verbenaceae, the verbena family < *verbena* Latin, a viburnum whose branches were used as a punitive flog and as a ceremonial whip < *verbera* Latin, lashes with a whip. The **werb* root is widespread in Indo-European languages where Latin compounds give current English words like *reverberate* to whip back and forth' and the Old Slavic word for willow tree, *vruba*, also a source of flogging switches, which has modern reflexes like *eeva* Russian, willow, and *vrba* Czech, willow. One important tree of the verbena family is *Tectona grandis* whose wood is teak lumber.

SPECIES

Lantana camara is the common, stiff-branched shrub grown as a pot plant or tender garden annual in northern gardens. The specific name is from *camara*, Spanish for room, chamber, hall, or anatomical cavity. *Camara* is a South American nickname for this plant which is an invasive, spreading weed in many tropical countries, growing into patches as big as a large room.

    Lantana montevidensis (Botanical Latin, from the vicinity of Montevideo, capital port city of Uruguay) is the trailing species used in hanging baskets and window boxes.

∾

LOBELIA

Dioscorides, author of "De materia medica"

Genus: Lobelia < named after Mathias de l'Obel (1538-1616), personal physician to England's King James I. A native of Flanders and an amateur botanist, de l'Obel helped popularize the plant when it was brought back in a British trading ship from the region around the Cape of Good Hope in South Africa. In 1570, he published an important early herbal under his scholarly Latin name, Lobelius.

Family: Campanulaceae, the campanula family, named after its typical genus < *campanula*, Late Latin, little bell < *campana* Late Latin, bell, referring to the shape of the flowers + *ula* Latin diminutive suffix, little

SPECIES

Lobelia erinus (Botanical Latin < *erinos* Greek, woolly; a name used by Dioscorides to describe an unrelated, basil-like plant. Pedanios Dioscorides who flourished in the middle of the first century A.D. was an army doctor, a Greek in the Roman army, who has some claim to having written the first, valuable pharmacopoeia. It was called *Materia Medica* and was the most copied and authoritative guide to the medicinal use of plants until the end of the Middle Ages.

A lovely white lobelia from America was introduced to European botany as early as 1665 beause it was falsely reported to have cured syphilis. *Lobelia syphilitica* (Botanical Latin, pertaining to syphilis) is still in hybridizers' greenhouses being used as a parent in modern crosses.

LUPIN OR LUPINE

Genus: Lupinus, the classical Latin name for the plant, supposedly < *lupus* Latin, wolf, in reference to the false notion that these plants wolfed down all the soil nutrients and left none for other flora. In fact, the nitrogen-fixing bacteria on the root nodules of legumes help improve the soil for other plants.

Family: Leguminosae, the legume or pea family, the second largest family of plants with more than 550 genera and at least 13,000 species < *leguminosus* Latin = *legumen* Latin, what one gathers or picks, notably peas and beans + *osus* Latin adjectival suffix indicating 'abounding in, full of.' *Legumen* < *legere* Latin, to gather (fruit), to select, to choose, to read (gather letters with the eyes). A large number of current English words derive from the basic Latin verb, *legere*, including collect, elect, lecture, legion, neglect, negligee, privilege, sacrilege, and select.

Legumes were first the fruit pods of peas and beans, the parts picked; then legume developed a later meaning of any vegetable grown for food. The legume family includes many foods like peas, beans, lentils, soybeans, peanuts, plus fodder and forage crops like alfalfa, clover, and vetch, as well as ornamentals like Wisteria and poinciana. The alternate, compound leaves of many leguminaceous genera are instantly recognizable.

SPECIES

Some of the commercial hybrids are derived from wild annual lupines like the Mediterranean *Lupinus luteus* (Latin, yellow-flowered), the Mexican *Lupinus pubescens* (Botanical Latin, lightly hairy, said of the seed pods), and *Lupinus hirsutus* (Latin, hairy, referring to the seed pods). *Lupinus texensis*, Bluebonnets, is one of the wild perennial forms and is the state flower of Texas.

Uses

The Romans grew one of the European annual lupines, *Lupinus albus* (Latin, white, of the flower) for oil extracted from its seeds and as fodder for domestic animals. In the 1930s, a British hybridizer named George Russell began crossing an American species, *Lupinus polyphyllus* (Botanical Latin, from Greek, many-leaved) with others like *Lupinus bicolor* (Botanical Latin, with flowers in two colours) to produce the famous Russell lupines still prized by gardeners of the temperate zone.

∾

MARIGOLD

Other common names: African Marigold, French Marigold (both from Central and South America—so much for the geographical accuracy of common names). John Gerard, Elizabethan herbalist, says these common names arose because the plant was introduced to Britain from France where the Holy Roman Emperor Charles V received marigold specimens as a gift from Hernándo Cortéz after he had sailed back to Spain in 1528 after conquering Mexico.

Several different plants are called marigold. Some Mediterranean calendulas are called pot marigolds, and the yellow-flowered *Caltha palustris* (Latin, of the swamp < *palus, paludis* Latin, swamp) is called Marsh Marigold.

Genus: Tagetes < *tageticus* Latin adjective, pertaining to Tages, an Etruscan god who sprang up from the earth like a young plant and taught men how to divine the future through reading natural signs like winds, thunder, lightning, and in particular by examining the entrails of sacred chickens, a prophetic practice the early Romans borrowed. The Etruscans spoke a language related to no other tongue so far discovered. The Romans who displaced them in central Italy borrowed some Etruscan words into Latin, among which were *amor* love, *persona* mask, *atrium* hall, and some Roman names like Cato, Cicero, Piso, and Varro. The names Roma, Remus, and so Romulus seem to have been taken from the Etruscan name of the original settlement on the Tiber. These interesting, mysterious, and cultured people also gave their name to Tuscany.

Family: Compositae, the daisy family < *compositus* Latin, placed together.

SPECIES

So much hybridization has occurred that true Tagetes species are seldom encountered in the trade. But there are about 40 species, including some offered by specialty greenhouses, such as *Tagetes tenuifolia* (Botanical Latin, thin-leaved, here finely divided) called signet marigold with tiny single flowers, and *Tagetes filifolia* (Botanical Latin, with thread-like leaves) grown for its filigree of lacy leaves and sometimes sold under the name Irish Lace.

WORD LORE

Marigold is a Middle English name for Tagetes from Mary Gold, where the Christian name refers to the Blessed Virgin Mary, mother of Jesus.

Uses

Widely used in North American gardens for their form and vivid colours, marigolds find a special use in Mexico, where the visitor sees vast hectares of orange-flowered marigolds grown to feed chickens. The fodder is the flowers themselves, eaten by hens which then lay eggs with the deep yellow yolks favoured by Mexican cooks. Planting certain marigolds in garden beds keeps down nematodes (parasitic worms). The Inca Marigold, *Tagetes minuta*, gives off a root chemical that kills or prevents many weed seeds from germinating in its immediate vicinity. Weeds like creeping charlie, bindweed, and couchgrass will not

Marigolds

grow close to these marigolds. A yellow dye is extracted from the roots of some species and used to colour paper. The tang of marigold petals balances the spicy sweetness of other flower petals in a potpourri.

Quotations

The Elizabethans who enjoyed the marigold in gardens took note of when its petals spread. In 1578, Henry Lyte wrote of the marigold in *A Niewe Herball* that "it hath pleasant, bright, and shining-yellow flowers, the which do close at the setting downe of the sunne, and do spread and open againe at the sunne rising." In *The Winter's Tale*, Shakespeare mentions "The Marigold that goes to bed wi' th' sun / And with him rises weeping."

☙

MORNING GLORY

Common names for various species: Cardinal Climber, Field Bindweed, Scarlet Star Glory, Spanish Flag

Genus: Ipomoea. Linnaeus coined this bit of Botanical Latin based on two Greek words, beginning with Theophrastos's term for a kind of wood-worm seen in vines, *ips, ipos*, and then adding *homoios* Greek, similar, like. The genus name referred to the curly, worm-like tendrils by which many morning glories climb and scramble over other plants.

Family: Convolvulaceae, the morning-glory family < *convolvulus* Latin, the word used by Pliny, the Roman encyclopedist, for the common European bindweed < *con (cum)* Latin, with + *volvere* Latin, to roll, in reference to the twining of the vine

SPECIES

Of the more than 400 species, *Ipomoea purpurea* (Latin, purple, of the flower colour) or *Ipomoea tricolor* is the parent of most annual morning glories. *Ipomoea batatas* is the sweet potato plant, with the species term from the Haitian name of the plant, the same Caribbean verbal root that gives the word *potato*. Several, spectacular, red-flowered Mexican species from *Ipomoea quamoclit* have a profusion of common names:

cardinal climber, crimson star glory, star ipomoea, Spanish Flag. Quamoclit is from a Central American language called Nahuatl, the tongue of the Aztecs, where it is *qua'mochitl* = *quaiutl* Nahuatl, tree + *mochitl* N., red vine.

~

Other common names of various species: Violet, Heart's Ease, Johnny-jump-up

PANSY

Genus: Viola < *viola* Latin, a sweet-smelling violet. The verbal root is a prehistoric, Mediterranean flower word related to *ion* Greek, a purple violet. An elaboration of the Greek root in Old French gives a name to the chemical element, iodine, because the gas and various salts were purple.

Family: Violaceae, the violet family

SPECIES
The most familiar of the 500 or so species are the tender perennial hybrids of *Viola wittrockiana*, everywhere grown as annuals. The species is named after the author of a once definitive monograph on the history of the cultivated pansy, a professor of botany named Veit Brecher Wittrock. *Viola tricolor* is Johnny-jump-up, one ancestor of the modern garden pansy. *Viola canadensis* (Botanical Latin, of Canada) is our wild violet that brightens woodland nooks early in the spring with white flowers tinged purple on the outside and yellow at the inside base .

WORD LORE
Pansy is an Englishing of the old French name, *pensée*, because the little nodding flowers seemed to mimic thoughtful faces. The nodding habit and gaudy colours of these flowers also

Violas or pansies

produced the derogatory slang use of pansy to refer to an effeminate, gay male.

❧

PERIWINKLE

Other common names: Creeping Myrtle, Running Myrtle

Genus: Vinca < *vincapervinca* Latin, periwinkle. This delightful, Roman, compound folk word appears frequently in Pliny the Elder and was borrowed directly into Old English as *perwince* well before the twelfth century. The basic root is *vincire* Latin, to bind, to fasten, to encircle, to wind around, so that *vinca per vinca* is playful street or children's Latin and is the Latin equivalent of an English phrase like loop-the-loop or bind-through-bind. This may refer to the Roman practice of making funeral wreaths from the long, sturdy stems of periwinkle by winding them round and round the circular base of the wreath, the base often made of osier or pliable wicker. Because the somber-green leaves do not wilt quickly after a stem is cut, periwinkle would have made a useful, persistent wreath.

Family: Apocynaceae, the dogbane family, named after its typical genus Apocynum < *apokynon* Greek, dogbane or a related plant < *apo* Greek, away, asunder + *kyon*, *kynos* Greek, dog; all from the belief that this plant could poison dogs. The dogbane family of 135 genera and more than 1,000 species is spread throughout the world, but is chiefly tropical. It includes ornamentals like frangipani, oleander, and allamanda, a woody greenhouse vine with perhaps the most formally beautiful yellow flowers on earth. Some genera of this family do have violently toxic juice, so powerful some Amazon peoples use it as arrow poison.

SPECIES

Vinca minor (Latin, smaller) is the common periwinkle used as a hardy, perennial groundcover because it flowers the first year, because the long, trailing stems root readily at nodes, and because periwinkle is one of the few hardy groundcovers for Canada that will thrive and bloom in the shade. It mats out quickly and is relatively unbothered by diseases and pests.

Vinca major (Latin, larger) is not so winter-hardy in Canada as *Vinca minor*. It is sold for window box planting as a tender annual, since it can be frost-killed.

Madagascar Periwinkle or Old Maid used to be in the Vinca genus, but has been reclassified and given its own genus and specific, *Catharanthus roseus* (Latin, pink, referring to the flowers). Catharanthus is Botanical Latin formed from two Greek words: *katharos* pure + *anthos* flower.

PETUNIA

Genus: Petunia, Botanical Latin < *petun* Obsolete French, tobacco < *petima* Tupi, tobacco plant. Tupi is the language of the Tupi peoples who live in the Amazon valley of Brazil. The showy little annual is in the same family, the nightshade family, and so closely related to the commercial tobacco plant and to Nicotiana or Flowering Tobacco. See our entry for Flowering Tobacco a few pages back in this discussion of the names of annuals.

Family: Solanaceae, the nightshade family < *solanum* Latin, the black nightshade plant. The nightshade family also includes such important food plants as potatoes, peppers, tomatoes, and eggplants. In this family are ornamental annuals like Browallia, Datura, Flowering Tobacco, Schizanthus, and poisonous plants like deadly nightshade.

SPECIES

Petunia x *hybrida*, the common petunia of commerce is a much-crossed species—hence the *x* preceding *hybrida*. F_1 hybrid petunias have flowers twice as big as hybrids and make vigorous plants. F_1 hybrids are made by cross-pollinating two different inbred hybrids. The problem is these F_1 hybrid produce seed that does not come true. So the gardener's price for big-blossomed petunias is the grower's boon: the gardener has to keep buying new petunias every spring because the F_1 hybrids will not reproduce exact copies of themselves from seed.

POPPY

Common names of different genera and species: Alpine Poppy, California Poppy, Iceland Poppy

Genus: Papaver < *papaver*, Latin, poppy. The word was borrowed into Old English as *popig* and by Middle English was *popi*. Both *opion*, the Greek word for poppy juice, and *opos*, Greek, vegetable juice, as well as *papaver* seem related to a Mediterranean root with a reflex in ancient Egyptian hieroglyphics as *peqer* 'poppy seed.'

Family: Papaveraceae, the poppy family, a large group of annual, biennial, and perennial herbs native to Eurasia, includes the opium poppy and various garden subjects.

SPECIES

Papaver alpinum, Alpine Poppy, is a sweet-scented dwarf annual from Europe's high mountains.

Papaver nudicaule (*nudus* Latin, bare + *caulis* Botanical Latin, stem < *kaulos* Greek, stem) is Iceland Poppy.

Eschscholzia californica is the California Poppy, whose generic name honours the first botanist to describe it, a Russian named Johann Friedrich Eschscholz (1793-1831).

WORD LORE

The Greek word for poppy juice is *opion* borrowed into Latin as *opium*. The opium of illegal commerce is obtained from the sap of unripe capsules of *Papaver somniferum* (Latin, sleep-inducing). This sap of the Opium Poppy contains the alkaloid, morphine. Heroin is formed by the acetylation of morphine. Opium is the oldest-known narcotic, mentioned as early as 4,000 B.C. in a Sumerian phrase that means 'joy plant.'

Poppy Day

Armistice Day, then Remembrance Day, in November recalls those who gave their lives in battle. Poppies were first sold by the British Legion to commemorate soldiers who fell in World War I. The association of the plant with the day arose from the vast fields of wild poppies on the battlegrounds of Flanders, and a Canadian poem helped cement this connection. "In Flanders

fields, the poppies blow / Between the crosses row on row"
was written by John McCrae, born in Guelph, Ontario, in 1872.
Major McCrae, First Brigade Surgeon, Canadian Field Artillery,
wrote the famous poem in 1915 during the second battle of
Ypres in Belgium. To great acclaim, it was published that year
in the British magazine, *Punch*. The poem is read during most
Remembrance Day services across Canada.

PORTULACA

Other common names: Rose Moss, Sun Moss, Sun Plant, Wax
Pink; and the pesky lawn weed, purslane or pigweed, belongs to
this genus.

Genus: Portulaca < Latin, purslane < *portula* Latin, little gate,
referring to the lid of its fruit capsule < *porta* Latin, gate. That
is one suggestion. More compelling is the etymology that says
portulaca was altered in Latin by folk change from some earlier
lost form like **porcellata*. *Porcellus* was a little pig, and the
stubby, succulent stems and leaves of this squat weed caused
people to call it 'the little pig.' Then some semi-literate Roman
heard *porcellata* as *porta lac* 'carries milk' because of the milky
sap of many species in this genus. *Lac* in Latin also denoted any
thick juice of plants. Some credence to this farfetched etymology
is given by the fact that in Late Latin *portulaca* actually did
show a variant *porcillago, porcillaginis* which was borrowed
into Old French as *porcelaine* and eventually gives the current
English word *purslane*. A twisty wordpath, but not an impossible
one.

Family: Portulacaceae, the purslane family, with 16 genera and
500 species

SPECIES

Portulaca grandiflora (Botanical Latin, big-flowered) is the
humble little moss rose that gives blossoms tinted in lovely,
papery pastels in poor hot soil where many other annuals will
mope and petulantly refuse to petal.

Portulaca oleracea (Botanical Latin, of the vegetable garden < *holus, holeris* Latin, potherbs, vegetables) is the piggy weed purslane or, in a British dialect variant that has much more pestiferous plosiveness, pussley. The specific epithet refers to the vegetable garden because some Europeans take revenge on this weedy intruder into gardens by eating the stems and leaves as salad greens.

∾

PRIMROSE

Primrose

Common names: Cowslip, Oxlip, Paigle

Genus: Primula < *primula veris* medieval Latin, literally 'first little thing of spring' < *primus* Latin, first + *ula* Latin feminine diminutive ending, little + *ver, veris* Latin, spring.

Family: Primulaceae, the primrose family

SPECIES

Of the more than 500 species in *Primula*, most are tender biennials and perennials grown as pot plants and put into Canadian gardens for the summer. But some will winter if protected. Hundreds of hybrids exist, but only one wild one will concern us here.

The Canadian primrose, *Primula mistassinica*, is now rare in the Pacific and western areas of its range. A hardy little primula that likes cool wet feet, it welcomes a May morning with a compact umbel of purply-pink flowers that have vivid yellow eyes. The plant was discovered in the spring of 1786 in the eastern part of its range in northern Québec. The year before, Louis XVI, monarch of France, had fallen into a regal funk because his gardens at Versailles had begun to bore him. "One more clump of *fleur-de-lys* and I'll eat my wig!" he may have pouted to France's greatest botanist of the day, André Michaux. So he shipped the obedient and delighted Michaux off to the new world to collect more inspiring specimens. One such gem Michaux found beside Lake Mistassini, named by the local Montagnais for the great stone that sat in the middle of the lake. *Mista-assini* means 'big-rock water' in Montagnais. Michaux called it a fairy primrose because it was smaller than the French species he knew.

WORD LORE

Primrose is an Englishing of *primula*. But the oldest word in our language for a species is cowslip which began in Anglo-Saxon or Old English as *cu-slyppe* 'cow shit,' such an unpleasant tag for so fetching an herb arising from the fact that these plants did well in a moist meadow near a plop of cow manure. *Slyppe* is cognate with *slipa* Old Norse, slime, dung, mud, and is related to *slop*. The origin of the British dialect name *paigle* is unknown.

Uses

Primrose flowers are made into a wine, used to perk up jams, and candied to decorate pastries. Young cowslip leaves add spice to salads and meat stuffings. Some people chop up the pungent root of *Primula vulgaris* (Latin, common) and add it to potpourri.

A False Accusation

A popular folk belief of gardeners is that all primroses cause skin irritation if one touches the leaves. No. Only varieties of the species *Primula obconica* do this, through glandular hairs on the underside of the leaves. The heads of these hairs contain a quinone called primin that will indeed irritate human skin but usually only in persons particularly prone to dermal allergies. But the stamens of many species can produce a mild, contact dermatitis. Don't touch the stamens and avoid vacuum-cleaner-like sniffings of the flowers.

Quotations

Shakespeare enjoyed the sound of the word, as in Act 2 of *Macbeth* where the porter's famous hell speech concludes,

"I had thought to have let in some of all professions that go the primrose way to the everlasting bonfire." In *Hamlet*, the chaste Ophelia advises her brother Laertes against "the primrose path of dalliance."

∾

SALVIA

Salvia or sage

Common names of some species: Cardinal Sage, Clary, Common Sage, Gentian Sage, Mealycup Sage, Scarlet Sage, Pineapple Sage, etc.

Genus: Salvia < *salvia* Latin, healing plant < *salvus* safe, saved, from the medicinal uses of several species. *Salvia* passed into Old French as *sauge* and thence into English as *sage*.

Family: Labiatae, the mint family < *labiatus* Botanical Latin, lipped, with a prominent lip < *labia* Latin, the lips. In botany, the labium is the lower lip of a flower with two lips. In flowers of the mint family, this labium is highly developed and enlarged, while the upper lip of the corolla though present is rudimentary. The mint family is large, with 224 genera and 5,600 species in the tropics and warmer temperate zones.

SPECIES

The only true annual in common use is *Salvia sclarea* (Botanical Latin variant of *clara* < *clarus* Latin, clear) which became clary or clear-eye in English, from the use of its seeds and leaves in eye lotions and ointments. Dozens of perennial salvias will bloom their first year and are sold as annuals.

Sage

Salvia officinalis is the herb used for centuries in sage and onion stuffings for meat or poultry. *Officinalis* 'of the shop' is a medieval Latin adjective seen as the specific name of many older botanical plants that were sold in shops because they had medicinal, food, or cosmetic uses. *Officina* is the medieval Latin word for a workshop or shop, from which English gets *office*.

∾

Other common name: Calf's Snout

SNAPDRAGON

Genus: Antirrhinum, Linnaeus's name < *anti* Greek, like + *rhis*, *rhinos* Greek, nose, snout of an animal. The flower looks like the snout of a dragon. Children love to pinch the dragon mouth (the laterals of the corolla) and make it snap open, hence the common English name.

Family: Scrophulariaceae, the figwort family, after its typical genus, *Scrophularia* which was named because herbalists thought it would cure scrofula, a now rare, tuberculous inflammation of lymph nodes and glands, once seen in children and adolescents. *Scrofula* is the Latin word for 'breeding sow.' The disease was so named because the inflamed glands in victims' necks made them look somewhat pig-necked. Also in this scrophulariacious family are annual foxglove (Digitalis), verbena, slipperwort, and monkey flower.

SPECIES

The much-hybridized garden annuals are offspring of *Antirrhinum majus* (Latin, bigger) and others of the thirty Mediterranean species.

∾

Genus: Helianthus < *helios* Greek, sun + *anthos* Greek, flower. This genus contains about 60 species, most native to North America.

SUNFLOWER

Family: Compositae, the daisy family < *compositus* Latin, placed together. This largest family of flowering plants is named after the compound flowers of its members. Small florets of individual flowers make up large clusters or heads.

Solis flos minior.

A somewhat fanciful sunflower with stars (!) from a 16th-century woodcut

SPECIES

Helianthus annuus (Latin, of a year), the common, giant, annual Sunflower is the state flower of Kansas, where nevertheless most farmers consider it a weed. Some of the other species have been crossed with *H. annuus* to produce the giant sunflowers of commerce. Many birds, especially finches, love the giant seeds that sunflowers produce in abundance. As for the common name, many sunflowers do well in light shade.

Jerusalem Artichoke or Canada Potato

In keeping with howlers among common names, this is neither an artichoke nor is it from Jerusalem. It is *Helianthus tuberosus* (Botanical Latin, with notable tubers or rhizomes, thickened rootstalks). Native to North America, this plant was also called Canada potato by early pioneers, and Girasole (Italian, turn-sun). Aboriginal peoples of North America cultivated the plant for its edible tubers, and white settlers in dire straits learned to eat it too.

∾

SWEET PEA

Genus: Lathyrus < *lathyros* Greek, one of the wild Mediterranean peas, probably < *lathon* Greek, hiding, hidden + *oura* Greek, tail, from the numerous tendrils by which peas scramble up and over anything in their way

Family: Leguminosae, the legume or pea family, the second largest family of plants with more than 550 genera and at least 13,000 species < *leguminosus* Latin = *legumen* Latin, what one gathers or picks, notably peas and beans + *osus* Latin adjectival suffix indicating 'abounding in, full of.' *Legumen* < *legere* Latin, to gather (fruit), to select, to choose, to read (gather letters with the eyes). Legumes were first the fruit pods of peas and beans, the parts picked; then legume developed a later meaning of any vegetable grown for food. The legume family includes many foods like peas, beans, lentils, soybeans, peanuts, plus fodder and forage crops like alfalfa, clover, and vetch, as well as ornamentals like broom, lupines, poinciana, and Wisteria.

SPECIES
These sweet-smelling peas are of the species *Lathyrus odoratus* (Latin, scented < *odor* Latin, smell) and make good climbers, borderers, bedders, and splendid cut flowers to brim a summer vase.

ॐ

VERBENA

See under *Family* in the entry for *Lantana.*

ॐ

ZINNIA

Genus: Zinnia < after Johann Gottfried Zinn (1727-1759), a professor of botany at the German university of Göttingen in Lower Saxony who successfully grew some seeds of what was later to be called *Zinnia peruviana* sent to him by a collector in South America.

Family: Compositae, the daisy family < *compositus* Latin, placed together.

Zinnia

SPECIES

The twenty or so species on which most modern hybrids are based are native to Mexico, Chile, Peru, Texas, and Colorado. *Zinnia elegans* is the most prolific parent of commercial cultivars.

BULBS

CORMS, RHIZOMES, & TUBERS

BULBS, CORMS, RHIZOMES, & TUBERS

DEFINITIONS & WORD LORE
OF BULB TERMS

All these plants have underground storage parts that are unlike normal roots. Inside its papery, protective tunic, a bulb contains a dormant, partially developed plant, together with enough food to maintain early growth.

Bulb

Bulbus Latin, any plant with a swollen root like hyacinth or onion < *bolbos* Greek, squill, grape hyacinth, narcissus, any bulbous plant < **bolbolos* Greek, reduplication of the Indo-European root **bhel* 'to swell up.' Distant relatives of bulb are phallus, phyllon (leaf), bull, and bubble.

A bulb is basically an underground stem surrounded by thick scales that are modified leaves containing food in the form of starches and sugars. The majority of bulbs evolved in and are native to the Middle East and southeastern Europe, where they enjoyed a moist spring in which leaves and flowers arose, followed by very dry summers when top growth died to the ground; but the bulb had stored food and water to enable the plant to

narcissus   ornithogallum

A crocus corm

bloom again the next spring. Hyacinths, narcissi, and tulips are such true bulbs. Each of their bulbs contains an embryonic stem and one or more embryonic "buds" from which new bulbs will develop during the next spring growing period.

Corm

Cormus Botanical Latin < *kormos* Greek, the trunk of a tree with its branches cut off < *keirein* Greek, to clip, to shear, to lop off

Unlike bulbs, corms are flat, as in gladiolus. Corms are the underground base of the stem swollen with starches and sugars on each side of the stem, stored there to provide nourishment for next year's growth. A true corm produces only one bud at its top. Crocus "bulbs" are actually corms. Unlike bulbs, corms shrink and wither after flowering, and new corms are formed on top of or alongside the old, withered corm.

Tuber

Tuber Latin, something swollen, a knot in a tree, a truffle < *tumere* Latin, to swell up. Related words are protuberant, tubercle, tuberculosis, tumescent, and tumour (*tumor* Latin, a swelling).

A cyclamen tuber

A tuber is the thickened terminal part of the stem used to store food. Begonia, Caladium, Cyclamen and Dog's Tooth Violet grow from such tubers. Dahlia is tuberous, but in its genus, the tuber is a swollen root. Tubers usually show more than one bud or eye, and they send up more than one growing stem. Many foods like potatoes and yams are starch-storing tubers.

Rhizome

Rhizoma Greek, the root mass of a tree < *rhizousthai* Greek, to take root < *rhiza* Greek, root, akin to *radix* Latin, root; to *root*, English, and to *wort* Old English, plant < Indo-European stem **ra*, to grow out of

A rhizome is a thickened root or underground stem by which the plant spreads, its rhizome sending out new flowering stems as it spreads, as in Lily of the Valley and Iris species. When the old part of a rhizome dies, a fresh, new extension is formed from which the next season's shoot will push up into the light. When you dig up bearded irises after two or three years, you can easily see the old, dead part of the rhizome, and at its end, firm new growth.

∾

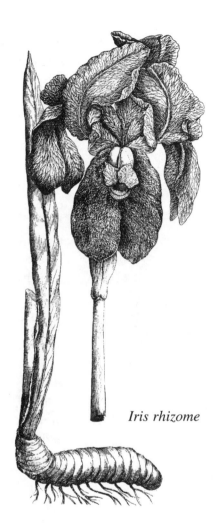

Iris rhizome

Genus: Allium < *allium* Latin, garlic < *all* Celtic adjective, hot, pungent, spicy. From the Latin comes the French word for garlic, *ail*. This large genus contains more than 300 species.

Family: Alliaceae, the capacious onion family, including chives, garlic, leek, shallots, and the ornamental onions like moly and giant allium. See *Herbs* section for onions in culinary use.

ALLIUM

SPECIES

For Canadian gardeners, the flowering onions fill a niche, blooming after most tulips and early spring bulbs have flowered, but before the midsummer show of lilies. One of our native wildlings that well stands transplanting to the garden is *Allium canadense* or Canada Onion. Its small flowerhead blooms in white, pink, or lilac shades. Showier but shorter is *Allium moly* (Greek, name of a yellow-flowered plant) which gives loose umbels of vibrant yellow in June. This has been a popular flowering onion since Elizabethan times. Shakespeare's contemporaries called these onions "mollies."

The Colossus of ornamental onions is *Allium giganteum* (Botanical Latin, unusually tall or large < *Gigantes* Greek, mythological giants so powerful they once laid siege to heaven. Zeus had to dispatch them with lightning bolts, and Hercules tidied up the surviving stray giants by clubbing them to death). These giant onions reach two metres in my raggedy Dunnville garden and bloom in spectacular, dense umbels of purply-blue. They are giants for the back of the border and need plenty of space. Fertilized liberally, giant onions reproduce with glee. I started with four bulbs about ten years ago, and now have sixty in the garden, and have given away many more. One bulb I left in place for four years had burgeoned into a most plump clump and produced twenty fat offsets by the time I dug it up.

WORD LORE

Onion < *oignon* French < *unio, unionis* Late Latin, a unity, a oneness of note < *unus* Latin, the numeral one. This word that gives modern English *union* also was used to name, e.g., a single, large pearl. Farmers adapted it to name a big, eating onion, a scrumptious oneness indeed, a delectable unity.

Uses

There are many reports of early uses of the wild Canada onion. The Blackfoot (Siksika) of Alberta made a tea of wild onion bulbs to soothe throats raw from coughing and to suppress the cough. Sinus congestion was treated on early prairie farms by inhaling smoke from a little bonfire of onion bulbs. All of the First Peoples of America had words to name wild onions. One of note is Cheyenne *kha-a-mot-ot-kewat* 'skunk testicles.' Eaten raw in too great a quantity, wild onion bulbs are toxic enough to cause gastroenteritis in young children.

AMARYLLIS

Genus: Hippeastrum, Botanical Latin < *hippos* Greek, horse + *aster* Greek, star. These tender bulbs for northern indoor pots are horse-stars because an early botanist thought the whole flowering plant of one species resembled a horse's head and the species was at that time called *Amaryllis equestre* (Latin, of a rider, equestrian). That species is now *Hippeastrum puniceum* (Botanical Latin, reddish-purple, named after Punic or Phoenician purple, a rich dye extracted from a Mediterranean mollusc and once used to provide the purple stripe that adorned the togas of noble Romans). There are more than 70 species in the genus, and many, many cultivars and hybrids based on plants native to Argentina, Brazil, the Chilean Andes, Mexico, Paraguay, Peru, and elsewhere in tropical America.

Family: Amaryllidaceae < *Amaryllis* Greek, a beautiful shepherdess in the pastorals of the Greek poet Theocritus, and borrowed into Latin poetry by Ovid and Virgil < *amaryssein* Greek, to have twinkling eyes. The English poets of the sixteenth and seventeenth century liked her name too, and had Amaryllises being rudely chased through silvan glades by sportive swains. Piano students may remember a ditty entitled "Amaryllis," said to have been composed by King Henry VIII.

SPECIES

The oldest species known to horticulture is *Hippeastrum puniceum*, the Barbados Lily, discovered in 1698. Since these big bulbs are so popular now in North America as winter pot plants, a brief history of their hybridization may be of interest. In 1799 a watchmaker in Prescot, England, named Arthur Johnson crossed *Hippeastrum vittatum* (Latin, striped lengthwise < *vitta* Latin, stripe) with *Hippeastrum reginae* (Latin, of the queen, named to honour Queen Caroline). A famous Dutch grower, de Graaf of Leiden, used the resulting *Hippeastrum johnsonii* as the base for a series of splendid amaryllises known as Dutch Hybrids, one of which, Appleblossom, bloomed this winter on the windowsill of the humble scriptorium where this book was written.

ANEMONE

See entry in the *Perennials* section.

BEGONIA

See entry in the *Annuals* section.

CALADIUM

Other common names: Angel's Wings, Elephant's Ears, Heart of Jesus

Genus: Caladium, Botanical Latin < *keladi* Malay, the name of an aroid with many-coloured leaves

Family: Araceae, the arum family < *aron* Greek, name of aroid plant related to our Jack-in-the-Pulpit. This family has 15 genera and more than 2,000 species, chiefly tropical. Most arums have arrow-shaped leaves and inconspicuous flowers. The most common foliage pot plant in the world is an arum, Philodendron. Also grown for ornamental foliage are Aglaonema (Chinese evergreen), Colocasia (also called Elephant's Ears), Dieffenbachia (Dumbcane), Pothos, and Scindapsus (Devil's Ivy, Golden Pothos, Silver Pothos).

SPECIES

Caladium bicolor (Botanical Latin, two-coloured) grows from starchy tubers planted in a moist, peaty loam indoors, then set out well after frost to bring exotic hues and startling leafiness to a poolside or patio for the summer. The fancy-leaved ones often belong to *Caladium hortulanum* (Botanical Latin, of gardens). *Hortulanus* is a cover-all specific usually indicating much hybridization and often the loss of the names of the specific parents.

Caladium hortulanum

Other common names: Camas, camass, commas, kamass, or quamash—there are more spellings for this once staple food bulb of western Canada than you can shake a digging stick at.

Genus: *Camassia* < *camas* Chinook Jargon (a trading *lingua franca* of nineteenth century Pacific coast use), name of the plant < *kamas* Nootka, sweet, of the edible bulbs. Tribes on the prairies used a western relative, *Camassia hyacinthina*, for the same alimentary purpose. The original Nootka name for the place that became Victoria on Vancouver Island was *Camosun*, 'place where we gather camas.'

Family: Hyacinthaceae, the hyacinth subfamily, belonging to the larger superfamily of lilies. See the *Hyacinth* entry in this section for the origin of the family name.

CAMAS LILY

SPECIES

Camassia quamash repeats the root in two forms. Once Pacific coast peoples used to harvest the bulbs of this blue-flowered member of the lily family and bake them immediately in ground ovens. They could be eaten hot, or dried and stored for winter rations. Another name for the plant was bear grass, because black bears would grub for the tasty bulbs in the summer. Humans however harvested them in the stout-bulbed autumn.

There is one fly in the paradisal ointment here—isn't there always?—and that is death camas, a nasty little plant that sometimes grows with camas and has bulbs similar in appearance, but it never has a blue flower. Death camas blooms a sickly white. *Zygadenus venenosus* is highly toxic to humans and other animals. Care had always to be taken at harvest. Indeed native peoples usually weeded out death camas when it flowered from among the food camas. In 1878 there was a camas war when the U.S. Army fought the Nez Percé people, after white settlers had let their pigs loose in the camas prairies that Nez Percé had used for centuries as natural gardens.

Several species including *Camassia cusickii* and the Canadian *Camassia quamash* make good garden subjects, planted like tulips in the fall and left undisturbed until overcrowding occurs. In its natural setting in British Columbia camas likes mountain meadows that are wet in the spring and that dry up well by midsummer.

CANNA

Canna Indica flore Luteo.

Canna indica

Other common names: Canna Lily, Indian Shot (British, referring to its quick, abundant growth)

Genus: Canna < *canna* Latin, reed, rush, cane < *kanna* Greek < *qanah* Aramaic < *qaneh* Hebrew < *ganu, gen, kenn* Egyptian variants < *gin* Sumerian, Akkadian. They all referred to cane plants like bulrush and slenderer reeds too. The English *reed* and *rush* are from Proto-Germanic words, while English *cane* derives through early French from Latin *canna*.

Family: Cannaceae, the canna family, of one genus, and about 50 species

SPECIES

Canna indica (Latin, here meaning of the West Indies) is the familiar yellow-flowered canna that almost explodes from its big tuberous rhizomes. Hybridizers working with species from Central and South America have produced variegated foliage and flowers in many colours besides the traditional red. But the big red spikes still look good against evergreens. Canna enjoyed a vogue as a cemetery planting in the American south in the middle of this century, which in turn caused red-flowered varieties to decline in popularity. New hybrids have seen them come back into favour with gardeners, even though they are not winter-hardy in most Canadian gardens and must be lifted every fall.

COLCHICUM

CAUTION: POISON

Other common names: Autumn Crocus, Meadow Saffron, Naked Ladies

All parts of autumn crocus are toxic. Colchicine is the chief alkaloid, most concentrated in the seeds and corms. Death has resulted from eating the bulbs as cures for gout and rheumatism. Children who have eaten the pretty flowers have died. Many people buy autumn crocus and let it bloom in a pot at a September window. I hope no kids or family pets think of nibbling.

Genus: Colchicum, named by Linnaeus < *colchicum* Latin, Pliny's word for meadow saffron < *Kolchis* Greek, Colchis, the name of an ancient city and of a country east of the Black Sea, which today includes parts of Armenia and Georgia. The ancient Greek name for autumn crocus was *kolchikon*. The Medea of Greek myth, a notorious maker of poisons, was from Colchis, and the ancient Egyptians and Greeks knew that the corms of autumn crocus were poisonous.

Family: Liliaceae, the lily family < *lilium* Latin, lily. The Latin word is related to *leirion* Greek, lily, and is one of the ancient Mediterranean flower words. Both Latin and Greek appear to have borrowed the word early from a Coptic form *hleli*, itself a variant of *hreri*, all stemming from ancient Egyptian *hrr*, lily.

SPECIES

Colchicum autumnale in its native Europe sends up leaves in the spring which then die down by midsummer. The plant goes dormant for two or three weeks, and then flowers in the autumn, with pale, mauve-pink blooms, which appear in long succession from early autumn sometimes right into early December. Colchicum is better in the ground as part of an alpine garden, where little explorers are less likely to eat any parts of this poisonous plant.

The giant corms of *Colchicum speciosum* (Botanical Latin, showy) produce equally giant flowers in a deep, crimson-purple. The biggest autumn crocus flowers (of course, it is not a crocus and not even related to the *Crocus* genus) belong to a variety of *speciosum* and they are white.

Colchicum luteum (Latin, yellow) is a native of the Himalayan foothills, and it blooms in early spring with small flowers of glowing yellow that sometimes leap up right out of the melting snow.

❧

Genus: Crocus < *crocus* Latin, saffron, a yellow dye and a spice obtained from the stigmas of a crocus < *krokos* Greek, saffron or its source, *Crocus sativus*. Theophrastos, an early Greek writer about plants, borrowed the name from a Chaldean

CROCUS

source, and so compare *karkom* Hebrew, saffron crocus, mentioned in the *Song of Solomon* in the Old Testament. The genus contains about 40 species, and hundreds of hybrids.

Family: Iridaceae, the iris family, a vast clan of 55 genera and more than 1,400 species, many gardenworthy. *Iris* is Latin from *iris, iridis*, Greek, rainbow, which was then personified as a goddess and charged by Zeus with carrying messages from the gods to humans across a polychrome sky bridge, the rainbow. Every time they saw a rainbow, the ancients thought a god was sending a private notice to some mortal.

SPECIES

Crocus sativus (Latin, cultivated, literally 'able to be sown or planted' < *serere* Latin, to sow) is the saffron crocus grown commercially now in Portugal and native to the Mediterranean littoral.

The yellow spice is made from the dried stigmas. It takes 500,000 crocus stigmas to make a kilogram of saffron, making it one of the more expensive spices. The etymology of saffron is < *saffran* Middle English < *safran* Old French < *safranum* Medieval Latin < *za'faran* Arabic, Persian, the yellow spice < *asfar* Arabic, yellow.

The giant-flowered Dutch hybrid crocuses are most popular, but do try some of the smaller species. Among the ones I have grown with great delight are *Crocus ancyrensis* (Botanical Latin, of Ankara), a species from Turkey with tiny, brilliant-orange flowers. It is often the very first spring bulb to bloom at my southern Ontario doorstep, popping through snowbanks late in a mild February or early in March. *Crocus etruscus* (Latin, of Tuscany, Etruscan) has a beautiful cultivar called Zwanenburg whose flowers are an unusual pastel blue. Finally there is my personal favourite of this genus, *Crocus susianus* (Botanical Latin, of Susa, in ancient Persia, now western Iran). This gaudy little native of the Crimea is also called Cloth-of-Gold Crocus for its March flowers of deep orange, striped with purple and set off by brilliant golden anthers.

Other common name: Sowbread (a sixteenth century loan translation of *panis porcinus* Medieval Latin, pig bread, or *Saubrot* German, sow bread, because the fat corms of cyclamen were a traditional European pig fodder)

Genus: Cyclamen < *cyclamen* Medieval Latin < *cyclaminos* Latin, Pliny's word < *kyklaminos* Greek, Theocritus's word for a Greek species, probably *Cyclamen graecum* < *kyklos* Greek, circle, referring to the fat, flat, round corms.

Family: Primulaceae, the primrose family, named after its typical genus, Primula < *primula veris* Medieval Latin, literally 'first little thing of spring' < *primus* Latin, first + *ula* Latin feminine diminutive ending, little + *ver*, *veris* Latin, spring. The family consists of 30 genera and 800 species.

SPECIES

Cyclamen persicum (Latin, of Persia) cultivars and strains provide the common Florist's Cyclamen for indoor growing. But some of the smaller species are winter-hardy in some Canadian gardens, if placed in a nook protected from the wind and well-mulched. Planted in late August, *Cyclamen coum* (Botanical Latin, from the Greek island of Cos) will bloom in late April. Put into the ground also in late summer, *Cyclamen europeum* and *Cyclamen neapolitanum* (Latin, of Naples) will bloom about fourteen months later, during the next autumn.

CYCLAMEN

Other common names: Adder's Tongue (Brit.), Fawn Lily, Trout Lily (all from the mottled dappling on the leaves of most species)

Genus: Erythronium, Latin < *erythronion* Greek, literally 'little red plant,' a name for some other red-flowered herb < *erythros* Greek, red. The Greek colour adjective has many derivatives in English technical vocabularies. For example, an erythrocyte is a red blood cell or red corpuscle, from *erythros* red + *kytos* Greek, cell. Erythromycin (*erythros* red + *mykes* mushroom + *in* common ending for organic chemical compounds) is an

DOG'S TOOTH VIOLET

antibiotic of wide application, named because it is produced by a red bacterium called *Streptomyces erythreus* that was originally isolated in a mushroom. Erythrism (*erythrismos* Greek, redness) is red-hairedness, of normal occurrence, with red hair, or beard, and a ruddy complexion.

Family: Liliaceae, the lily family < *lilium* Latin, lily

SPECIES

One European species widely planted in the early autumn in North America is the true Dog's Tooth Violet, *Erythronium dens-canis* (Latin, tooth of the dog), so named because its conical little corms resemble a canine incisor. But North American species have wintered over with more success in my modest plot of ground than the big European poop which has been winter-killed frequently whilst under my apparently oafish ministrations. *Erythronium americanum* or Yellow Adder's Tongue has survived for me. Gardeners out west have better luck with native species like *Erythronium oregonum* (Botanical Latin, of Oregon), White Fawn Lily, and *Erythronium californicum.*

Dog's Tooth Violet, Fawn Lily, or Trout Lily

ERANTHIS

Other common name: Winter Aconite

Genus: Eranthis, Greek < *er, eros* Greek, springtime, cognate with *ver*, Latin, spring + *anthos* Greek, flower. The six tuberous species are native to Europe and are among the earliest to bloom of all spring bulbs.

Family: Ranunculaceae, the buttercup family, named after its typical species, the buttercup, Ranunculus < *rana* Latin, frog + *unculus* Latin, diminutive suffix, little, tiny. *Ranunculus* is Latin for little frog, so named because buttercups like damp places just as amphibians do.

SPECIES

Eranthis hyemalis (Latin, of the winter < *hiems, hiemis* Latin, winter, cognate with *chion* Greek, snow. See chionadoxa later in this section under Glory of the Snow). This eranthis blooms in January in its native Europe, with large, pale yellow, buttercup-like flowers. From Cilicia, an ancient country in what is now southern Turkey came *Eranthis cilicia* with finely cut, bronzed leaves and deep yellow petals. Van Tubergen, the famous Dutch bulb nursery of Haarlem, crossed *E. cilicia* with *E. hyemalis* to produce a very large, golden-hued eranthis with sterile flowers that last longer than fertile flowers, often blooming for a month. It is sold as *Eranthis tubergenii*.

 ❧

Other common names: Desert Candle, Giant Asphodel

FOXTAIL LILY

Genus: Eremurus, Botanical Latin < *eremia* Greek, a solitude, a deserted place, a desert < *eremos* Greek, lonely, solitary ı *oura* Greek, an animal's tail. The single, tall spike of the inflorescence suggests the tail of a fox on these giant lilies that are native to the deserts and steppes of Afghanistan and northern Iran in central Asia. One who dwells alone in the desert in Greek was an *eremites*, which evolved into English *hermit*.

Family: Liliaceae, the lily family < *lilium* Latin, lily. The Latin word is related to *leirion* Greek, lily, and is one of the ancient Mediterranean flower words. Both Latin and Greek appear to have borrowed the word early from a Coptic form *hleli*, itself a variant of *hreri*, all stemming from ancient Egyptian *hrr*, lily.

SPECIES

Startling garden plants one to four metres high result if you can get them to overwinter and to bloom. Springing out of a bundle of sword-like leaves is a tall stem bearing hundreds of yellow, pink, or orange bells early in the summer. The brittle tubers—handle with care—are put into well-drained loam about 5 centimeters deep in September. Mark the site well, because eremurus may not bloom the first year. But when they do—Wow! Mulch in early spring with leaves to save tender early growth from frostkill. If there is no snow cover during your winter, protect soil over the tubers with evergreen branches. This fussbudget is worth all the trouble.

Available and expensive species include *Eremurus bungei* (Botanical Latin, after Alexander von Bunge, 1803-1890, Estonian botanist who wrote a monograph on the foxtail lilies), one metre tall, with golden-yellow spikes, the hardiest foxtail lily in Canadian gardens. *Eremurus himalaicus* (Botanical Latin, of the Himalayan Mountains) has white flowers; well-grown specimen plants have attained heights of four metres in southern Canada. *Eremurus robustus* can reach two metres and bears pink blooms. British hybridists have produced a range of flower colours, for example the "Shelford Hybrids" appear in white, orange, yellow, cream, rose, and peach-pink.

FRITILLARY

Common names of various species: Checkered Fritillary, Chocolate Lily, Crown Imperial, Guinea-Hen Flower, Mission Bell, Northern Rice-root, Snake's Head

Genus: Fritillaria, Botanical Latin < *fritillus* Latin, a dice cup in which dice were shaken. Fancy Roman dice cups were wooden, with inlaid designs of marquetry done in a chessboard pattern, much like the markings on the corollas of *Fritillaria meleagris*, Checkered Fritillary. But the shape of the seed capsule or even the shape of the flower may have reminded an early botanist of a dice cup. Several species of butterfly with checkered wing patterns are also called fritillaries.

Family: Liliaceae, the lily family < *lilium* Latin, lily

SPECIES
Crown Imperial, *Fritillaria imperialis* (Botanical Latin, like an emperor, showy, exotic) does have the air of a regal headdress

Corona Imperialis alta Rubicunda.

Corona Imperialis Coronata flauo flo.

Crown Imperial or
Fritillaria imperialis

with a top tuft of lance-like leaves protecting the drooping but
vivid flowers beneath, which hang in thick clusters in shades of
reddish orange, bronze, yellow, and bright red. The big bulbs and
the glorious flowers smell bad. The bulbs smell so bad that they
will keep mice and squirrels away from any patch. Plant crown
imperials well away from windows and areas where you sit outdoors. On the other hand, don't avoid these pricey bulbs either.
The smell, though rank, does not take over a bed, and is easily
disguised by the more pleasing aromas of nearby plantings.
Carolus Clusius (the humanist name of Charles de l'Écluse,
1526-1609, a famous Flemish botanist) introduced crown
imperials into Europe from their native stands in Turkey and
Persia.

Also popular in North American gardens is Checkered Lily,
Fritillaria meleagris (Botanical Latin, with spotted colouring,
like the feathers of the African guineafowl, *Numida meleagris*).
In Greek mythology, Meleager was a hero. When he died, his
sisters mourned so plaintively at his death that Artemis changed
them into guineahens. It is said Artemis did this out of pity. But,
I wonder, could she have been annoyed rather at their cooing
and moping, and vindictively decided they deserved to make
such sounds for all eternity?

Several species are native to western Canada and will thrive
in Pacificside gardens. *Fritillaria Camschatcensis* (Botanical
Latin, of the Kamchatka peninsula at the extreme eastern end of
Russia, between the Bering Sea and the Sea of Ohkotsk) is also
found on the other side of the Pacific Ocean, from Alaska, down
through coastal British Columbia, with abundant stands on the
Queen Charlotte Islands, to the limit of their southern range in
the state of Washington. The curious bulbs look like
a bunch of grains of cooked rice, hence one common name,
Northern Rice-root. The flowers hang like dark-purple bells,
hence another common name, Black Lily. Southern Inuit and
many First Peoples of the Pacific coast dug the starchy bulbs in
the autumn, dried them, and powdered them to use as a winter
flour.

Fritillaria lanceolata (Botanical Latin, with lance-like
leaves) is also called Chocolate Lily from its strange, purplish-
brown flowers mottled with green-yellow splotches. It grows
wild on southern Vancouver Island.

Another species to be enjoyed, but not to be picked in the wild, is *Fritillaria pudica* (Latin, ashamed < *pudor* Latin, shame) which hangs its little, nodding, yellow flowers like a bashful beau. You can almost see this modest little wildling toe the dirt and mutter, "Aw, shucks." It has common names like Mission Bell in its U.S. range, and Yellow Bell in Canada.

ॐ

One, now rare, common name: Sword Lily

Genus: Gladiolus, Pliny's term for one species of Sword Lily < *gladiolus* Latin, a small, short sword, diminutive form of *gladius* Latin, the most common military sword of the Romans, and the one swung in the public games by swordsmen called *gladiatores*. The plant name refers to the long, sword-like leaves of most species. The Latin plural form is usual in English too: gladioli.

Family: Iridaceae, the iris family, a vast clan of 55 genera and more than 1400 species, many gardenworthy. *Iris* is Latin from *iris, iridis*, Greek, rainbow.

SPECIES

Plant breeders crossed about 12 of the more than 150 wild species in this genus to produce *Gladiolus hybridus*, the "glads" of modern horticulture. The thousands of current cultivars have such an intertwined ancestry that gladioli are classed by colour and flower size. The corms are usually lifted in the fall in Canadian gardens, although some gardeners in southern Canada have reasonable luck burying them deep and getting three or four overwinterings.

ॐ

Genus: Chionadoxa, Botanical Latin < *chion* Greek, snow + *doxa* Greek, glory, praise, opinion. The generic and common name refer to its habit of early spring bloom, often pushing up through the last, vestigial snow patches. Related words in English include *doxology* through medieval Latin from *doxologia* Late Greek, literally 'writing that praises (God).'

GLADIOLUS

GLORY OF THE SNOW

In the Anglican Church, the hymn "Praise God from whom all blessings flow" is popularly referred to as the Doxology. The Greater Doxology is the hymn *Gloria in Exclesis Deo*, part of Holy Communion. In its prime meaning, something of correct or true or straight religious opinion was orthodox (*orthos* Greek, straight, true + *doxa* opinion). Something of another opinion was heterodox (*heteros* Greek, different); something of contrary opinion was paradoxical (*para* Greek, beside, away from normal). From the same verbal root, *dokein* Greek, to believe, to seem, came religious belief or dogma.

Family: Liliaceae, the lily family < *lilium* Latin, lily

SPECIES

Chionadoxa luciliae, the type species, is still widely planted in North America and Europe. A Swiss botanist from Geneva, Pierre Edmund Boissier (1810-1885) was collecting specimens in the Tmolus Mountains in western Turkey, named after King Tmolus of ancient Lydia in Asia Minor. High up in the melting spring snow of one range, Boissier found the violet-blue, white-centred flowers, blooming at the edge of retreating snowcover and named them after his wife, Lucile.

∾

GRAPE HYACINTH

Genus: Muscari. The famous Flemish botanist, Clusius (Charles de l'Écluse, 1526-1609) was the first European to make note of the name, on bulbs he received from Turkey in 1583. They had been labelled *Muscari*, *Muschorimi*, *Muschoromi*, *Muschio greco*. From this welter of Arabic (*romi*, Roman), Turkish (*rimi*, Roman), Italian, Latin, and Greek emerges the basic root for musk, referring specifically to the sharp scent of *Muscari moschatum*, the musk hyacinth of Asia Minor, which gave its name to the genus. Musk < *muscus* Late Latin < *moschos* Late Greek < *mushk* Persian, musk, a perfume obtained from the male musk deer, *Moschus moschifer* < *muska* Sanskrit, vulva, scrotum < *mus* Sanskrit, mouse < **mus* Indo-European root, mouse.

The Sanskrit *mus* has a derivative *muska* literally 'the mouse place of the human body, the crotch where pubic hair has a mouse-like appearance,' then the meaning expands to refer directly to the scrotum or the vulva. Old Persian borrowed that Sanskrit word to name an odorous extract from the sac under the abdomen of the musk deer.

Family: Hyacinthaceae, a subfamily of the lilies, named after its typical genus, Hyacinthus. See the next entry for origin of the name.

SPECIES

The common name of grape hyacinth was first applied to the racemes of almost black little flowers clustered on each stem of *Muscari botryoides* (Botanical Latin, resembling a bunch of grapes < *botrys* Greek, bunch of grapes). They even have a powdery bloom like grapes. John Ruskin (1819-1900) , the English art critic and man of letters, was impressed on seeing his first Muscari and wrote that it was "as if a cluster of grapes and a hive of honey had been distilled and pressed together in one small boss of celled and beaded blue."

Grape hyacinths are sturdy, dependable little bulbs, winter-hardy, spreading nicely over the years, and giving deep blue and purple underquilting for the larger early tulips and daffodils. The stars for me are known in British commerce as Oxford or Cambridge grape hyacinth, *Muscari tubergenianum* (Botanical Latin, of Van Tubergen, the Dutch bulb nursery of Haarlem, who introduced this exquisite turqoise-blue-flowered alpine from the mountains of northwest Iran).

An interesting oddity is *Muscari comosum* var. *monstrosum*, the plumed grape hyacinth, in which the flowers are reduced to thin, filamentary tassels, making the flowerheads look like an insane spider trying to turn into a feather. These little weirdos last long in bloom and make startling additions to a spring vase of cut flowers.

HYACINTH

Genus: Hyacinthus, Latin < *Hyakinthos* Greek, a flower, a gemstone, and the name of one of Apollo's handsome boyfriends < apparently related to *Bakinthos*, name of a spring month on the island of Crete. Hyacinth is not a Greek word, but appears to be of ancient Aegean provenance. All the Greek place names in *-nthos* existed before the Dorians invaded the

Hyacinth

Hellenic peninsula, bringing the Greek language. In Greek mythology, Apollo and Zephyros, the West Wind, were both smitten by the beauty of Hyacinth and they constantly fought over which of them should be his lover. Then, one day while Apollo was teaching Hyacinth to throw the discus, Zephyros, in a jealous snit, blew the thrown discus by a mighty blast of wind back into Hyacinth's face, smashing his beauty and killing the boy. As a memorial, Apollo caused the flower to spring forth from the boy's spilled blood. The Greeks also thought one could see, on each petal of one particular Greek species, written in capital letters the Greek exclamation of woe, *AI*. This myth is the garbled remembrance of a spring vegetation deity who rose in the spring, died by summer, and was reborn again the next spring.

Family: Hyacinthaceae, a subfamily of the lilies, named after its typical genus, Hyacinthus.

SPECIES

There are about 25 species in the genus, which grow around the Mediterranean, native to places like the Pyrenees in Spain, the island of Corsica, Dalmatia in the former Yugoslavia, through to Iran. Some gardeners enjoy what Dutch hybridizers have done to species like *Hyacinthus orientalis* (Latin, of the east, in this case, native to Turkey, Iran) on which most commercial hyacinths are based. They have taken a modestly floriferous beauty and transformed it into a gaudy spike jammed with florets and oozing a smell so rich and powerful it might knock over a camel. The common hyacinths of autumn pot culture and bedding now look like something that might decorate a rococo brothel. Although they are seldom offered for sale nowadays, having being pushed aside by the fat Dutch hybrids, the patient searcher can sometimes discover the true species hyacinths, with spikes of fewer florets. One to seek is *Hyacinthus amethystinus* (Botanical Latin, of amethyst blue), a charming plant available in clear blue or white forms.

See entry in the *Perennials* section.

IRIS

❧

Common names: Hundreds of plants not in Liliaceae have *lily* names, like Trout Lily. These confusing nicknames should be discouraged.

LILY

Genus: Lilium, Latin. See below for etymology. The genus contains about 80 species, widely distributed in the northern hemispheres of both the Old and the New World. Because they make such generally carefree and lasting contributions to gardens, lilies have been much dickered with and hybridizers have produced thousands of cultivars.

Family: Liliaceae, the lily family < *lilium* Latin, lily. The Latin word is related to *leirion* Greek, lily, and is one of the ancient Mediterranean flower words. Both Latin and Greek appear to have borrowed the word early from a Coptic form *hleli*, itself a variant of *hreri*, all stemming from ancient Egyptian *hrr*, lily. The lily family is so huge that subfamilies have been formed. For example, the day lily has its own little group here called Hemerocallidaceae, for which see the *Day Lily* entry in the *Perennials* section.

SPECIES

Of the thousands of lilies, we have space here in this little printed garden for some lilies native to Canadian woodlands, many of which do well under cultivation. Canada Lily, *Lilium canadense*, was called Meadow Lily by pioneers. Dismissed by snobs as Ditch Lily, this sturdy little bloomer offers cup flowers that range from deep yellow through orange, with black or brown spots, on stalks just above one metre tall.

Lilium philadelphicum (Botanical Latin, of Philadelphia, the city in Pennsylvania) was named by Linnaeus after bulbs were sent to the Swedish botanist by his pupil Kalm, who was collecting North American species. This red-to-orange beauty is commonly called Prairie Lily, Western Red Lily, and Wood Lily, and is the floral emblem of the province of Saskatchewan.

British Columbia has a native lily throughout the southern half of the province. *Lilium columbianum* (Botanical Latin, of the Columbia River area) is, interestingly, not very common in the dry valley of the Columbia River. It has orange flowers spotted maroon. A common name is Tiger Lily.

The Tiger Lily In Korean Folktales

The Korean myth about how the Tiger Lily received its name is worth repeating, showing, as it does, the worldwide spread of certain stories. Read the Korean myth, and then think of Androcles and the Lion. One day, a Korean woodsman, living as a hermit deep in a lonely forest, happened upon a young tiger, lying wounded in the camouflage of some underbrush. The big cat had been shot with an arrow that still stuck out of its foreleg. The woodsman removed the shaft, nursed the tiger back to health, and the cat became devoted to the woodsman. But even into this dream of the peaceable kingdom came death. Years later, before the tiger died, the beast pleaded with the woodsman to use his magic so that the spirit of the cat could be beside him always. The tiger died. After many magic incantations, the woodsman succeeded in turning the body of the beast into the first Tiger Lily. A few years later, the woodsman, grown old, was crossing a stream in the full torrent of spring run-off, when he slipped on a mossy log and drowned. Back beside the humble hovel deep in the forest, the Tiger Lily looked for his friend day after empty day. Then the Tiger Lily decided he would have to spread, as only a plant can, over all of Korea, and, as he spread, the Tiger Lily would keep looking for his lost companion, turning his bright face to the summer for all the years to come.

Lilium candidum (Latin, shining white—a Roman candidate wore a white toga) is the Madonna Lily, an introduced species, and quite difficult to grow in Canadian gardens. Nevertheless, it is the official floral emblem of Québec since 1963, because it resembles the *fleur-de-lis* (archaic spelling: *fleur-de-lys*), a common French heraldic device which adorns armorial bearings and Québec's flag. Wonderful indeed. Of course, the fleur-de-lis was not a lily. The word *lys* in fact has nothing whatsoever to do with lily or *lilium*. *Fleur-de-lys* means in Old French 'flower native to the Lys River in France.' Old French river names predate the French language, probably by millennia. Lys may be

related to Celtic *li* 'white,' a reference to the freshness of the
river's water, to shallows and rapids in the river, or to chalky
soil formations along its banks. No trace of lilies native to
its banks has been found. The fleur-de-lis was an iris, *Iris
pseudacorus*. Québécois botanists have suggested that a better
floral emblem might be one of our native Canadian irises, the
Blue Flag, *Iris versicolor*. It grows widely in Québec, but is of
course tainted by the fact that it is native also to other parts of
Canada. Perhaps someone will happen upon a blossom free of
Canuck taint, mayhap a wee posy from Mars?

Quotations
"Consider the lilies of the field, how they grow; they toil not,
neither do they spin: and yet I say unto you, that even Solomon
in all his glory was not arrayed like one of these." That beautiful
metaphor from the Gospel according to Saint Matthew 6: 28-29
is justly famous. Originally written in Koine Greek "lilies" is a
translation of *ta krina*, probably not lilies at all. Ancient
Palestine had few species of Liliaceae. The flower mentioned is
likely to have been the bright scarlet *Anemone coronaria*, which
was abundant in the Holy Land. See under *Anemone* in the
Perennials section.

∾

Obsolete seventeenth-century English common name:
Convally

LILY OF THE VALLEY

Genus: Convallaria, Botanical Latin < *lilium convallium*,
Medieval Latin, herbalists' name for lily of the valley
< *convallis* Classical Latin, a valley enclosed on all sides
< *vallum* Latin, an earthen wall set with stakes, a palisade, a
rampart < *vallus* Latin, stake, pale. English *valley* derives from
French *vallée*; English *vale* from Old French *val*; both French
forms < *vallis, valles* Medieval Latin, valley.

Lilium convallium was borrowed from the phrase used by
St. Jerome about A.D. 384-404 when he was making a Latin
translation of the Bible, to establish a standard edition
sanctioned by the Roman Catholic Church. The name of this
translation of the Bible was *editio vulgata* 'common edition.' It

is known in English as the Vulgate. St. Jerome used *lilium convallium* to translate the Hebrew of part of the opening verse of chapter two of *The Song of Solomon*, which much later would be rendered in the King James's English version of the Bible as "I am the rose of Sharon and the lily of the valleys."

Family: Convallariaceae, the lily of the valley family, a subfamily of the lilies, with one genus and one species

SPECIES

Lily of the valley is *Convallaria majalis* (Botanical Latin, flowering in May < *mensus Maius* Latin, the month of May, named after the ancient Italic earth goddess Maia < *maior* Latin, greater, so that May is the month of increase, the month of greater and renewed growth). Every gardener should consider this delight which increases by a creeping, underground rhizome that develops crowns or pips. The small pips send up long-lasting leaves; the big pips sprout flower stalks that bear little, bell-like nodding flowers that give off a sweet, spring perfume. Here is an easy-to-bloom perennial that actually prefers to be in the shade of trees and shrubs! Some books brand lily of the valley as invasive. But it takes four or five years before the rhizomes creep too far in their determined carpeting of a garden patch. Note that no parts of lily of the valley are to be eaten, as the plant contains more than twenty cardioactive glycosides, including convallotoxin and convalloside.

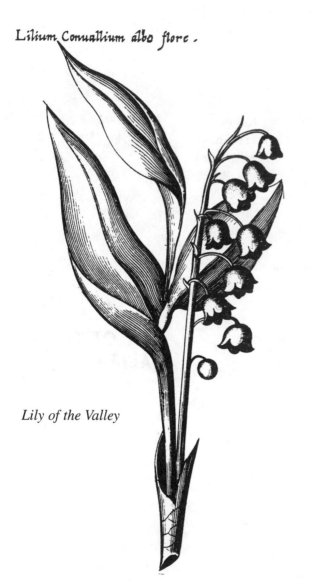

Lilium Convallium albo flore.

Lily of the Valley

Common names of various species: Daffodil, Jonquil, Lent Lily, Paperwhite

NARCISSUS

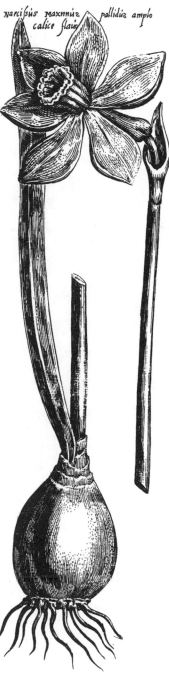

narcissus maxnnuz
calice flauo   pallidus amplo

Genus: Narcissus, named by Linnaeus from a figure in Greek mythology < *Narkissos* Greek, a beautiful boy in the old myths.

Narkissos was so handsome that he spurned the love of women and men. But one day beside a forest pool, the lad found his true love—himself—while gazing into the water. Narkissos kept reaching to grasp his own reflection in the water, but each time he touched the surface, his image disappeared. He could not bring himself to leave the bank of the pool and so he stayed there, forgetting to eat, entranced by his own beauty. When he eventually expired, the gods turned him into a golden flower.

The ancients thought that his name was related to a Greek word for stupor or stunnedness, *narke*, a word that is the source of English words like narcolepsy, narcotic, and Nark Squad. In this reading of his name, Narkissos was "stunned" by his own gorgeousness. But Greek personal names and place names in *-issos* were borrowed from other languages, as *-issos* is not of Hellenic origin. It seems likely this mythical figure first arose in ancient Persian, and the Greeks borrowed the Persian name, and by folk etymology made it look as if it were related to *narke*.

Havelock-Ellis and Sigmund Freud used narcissism to label an early stage of sexual development wherein our own bodies arouse erotic feelings in us. According to Freud, most of us outgrow this fixated period of intense self-love. Freud, of course, never sat beside the runway at a fashion show, where self-love gets so revved up that some of the more fastidious couturiers have taken to passing out plastic splash guards to attendees sitting in the front rows. These guards, however, do not prevent customers from overhearing the model who cooes to a mirror, "God, I'm so scrumptious, nobody gets me— man, woman, or beast. Well—maybe Lassie, but only if she begs."

The genus *Narcissus* comprises about forty, mostly European species. The horticultural forms have been crossed and recrossed for hundreds of years, and now number in the thousands of varieties and cultivars.

Family: Amaryllidaceae, the amaryllis family < *Amaryllis* Greek, a beautiful shepherdess in the pastorals of the third-century B.C. Greek poet Theocritus, and borrowed into Latin poetry by Ovid and Virgil < *amaryssein* Greek, to have twinkling eyes

SPECIES

The many, beautiful hybrids are widely available, but do look out for some of the smaller Narcissus species offered for sale from time to time. Among my favourites is *Narcissus poeticus* or Pheasant's Eye Narcissus with very fragrant, wavy-edged, white petals thinly margined in deep red. *Narcissus minimus* (Latin, very small) is among the smallest bulbed plants in the world, native to Asturia in the Spanish Pyrenees. One cultivar "Jack Snipe" is available early in the fall at garden centres. Try a dozen of these tiny beauties tucked into a nook beside your front door.

WORD LORE OF DAFFODIL AND JONQUIL

Asphodel, the name of a narcissus frequently mentioned in ancient poetry, derived from *asphodelos* Greek, a lily. But plant names get tossed around in history as they are borrowed from language to language. "The asphodel" in Middle Dutch was *de affodil*, borrowed into English to produce daffodil!

Jonquil entered English from Early French *jonquille*, *Narcissus jonquilla*. French borrowed the word from *junquilla* Spanish, little reed, named because of the rush-like leaves of many narcissi species. The Spanish word goes back to *iuncus* Latin, rush, reed.

∾

OXALIS

Other common names: Wood sorrel

Genus: Oxalis, Botanical Latin < *oxys* Greek, sharp, sour-tasting, acidic, named because the leaves of many sorrels have a slightly bitter taste, but were used to make meat and fish sauces.

This is a large genus with more than 800 species, most tuberous, native chiefly to South America and southern Africa, but with several European and North American Wood Sorrels in the family too.

Family: Oxalidaceae, the oxalis family, formerly included in Geraniaceae

SPECIES
A common garden subject is *Oxalis adenophylla* (Botanical Latin, with glands on the leaves) from Chile. It blooms with pink and white flowers from a little fan of greyish leaves. European Wood Sorrel is *Oxalis acetosella* (Late Latin, name for a plant with acidic leaves).

WORD LORE OF SORREL
The sorrel of Elizabethan cookery came into English from Medieval French *surele* (*surelle*, *surette*, Modern French), itself derived from *sur* Old French, sour < *sur* Old Germanic, sour; compare modern German, *sauer*.

*Squill
or Scilla*

Other common name:
Squill

Genus: Scilla, Botanical Latin < *skilla* Greek, a different plant, the sea onion, *Urginea maritima*. The English *squill* is influenced by confusion with *squilla* Latin, a crustacean, but the plant name appears to be of broad Indo-European origin because there is a Proto-Germanic form, *skilla*, not borrowed from Latin or Greek.

SCILLA

Family: Hyacinthaceae, a subfamily of the lilies, named after its typical genus, Hyacinthus. See *Hyacinth* in this section for the etymology of the name.

SPECIES

Of the roughly 80 species in this genus, several are widely available, including the best-known member of the genus, *Scilla sibirica* (Botanical Latin, of Sibiria). Drooping flowers in two shades of deep blue make this early bloomer welcome in any bulb garden. It spreads rapidly, merrily winding its way through other bulb plantings. One useful cultivar is "Spring Beauty" with sterile flowers that consequently last for two weeks or more.

❧

SNOWDROP

Other common names: Milkflower, *Pierce-Neige*

Genus: Galanthus, Botanical Latin < *gala* Greek, milk + *anthos* Greek, flower, from the white flowers of many species

Family: Amaryllidaceae, the amaryllis family < *Amaryllis* Greek, a beautiful shepherdess in the pastorals of the third-century B.C. Greek poet Theocritus, and borrowed into Latin poetry by Ovid and Virgil < *amaryssein* Greek, to have twinkling eyes

SPECIES

About twenty species make up this genus of minor bulbs that are nevertheless perfect for any Canadian garden site that is cold and bleak. Snowdrops hate heat. They can also be potted up in September and forced into bloom to make a pleasant addition to a Christmas table. *Galanthus nivalis* (Latin, snow-white or growing near snow) is one usually available late in August, and it has many cultivars that are outstanding additions to the cold garden.

WORD LORE

Snowdrop entered Elizabethan English as a direct translation of one of several German folk names, probably *Schneetröpfchen* 'snow droplet,' named because of the semi-globular shape of half-open flowers, and because this early spring bloomer will poke through melting snowdrifts in deciduous woodlands of its native Europe.

∾

Genus: Tulipa Botanical Latin < *tulipa* Early Modern Dutch < *tulipan*, *tolliban* Early French < *tülbend lale* Colloquial Turkish, the turban white, name of one species native to Turkey < *tulbend* Turkish, turban, because the shape of the flower

TULIP

An Elizabethan Tulip

resembles the Turkish headdress and the flower colours are vivid like the colours of the fabrics used to make turbans. Long before they saw the living flower, Europeans saw the tulip as a motif in Turkish ceramics. The bulbs had been grown for centuries in Anatolian gardens. Tulips were introduced into European commerce in the year 1554 by a man with the wonderful name of Ogier Ghiselin de Busbecq, who was Ambassador of the Holy Roman Empire to the sumptuous court of Suleiman the Magnificent, the Ottoman sultan under whose expansionist rule the Ottoman empire reached its zenith. Busy de Busbecq, busy as a bee, first saw tulips on his way to bow before Suleiman in Constantinople. Soon they were the plant rage of the continent, attaining so frenzied a popularity by 1634 in Holland that a madness actually called tulipomania resulted. Prices for single, rare tulip bulbs rose in speculative hysteria to 100,000 florins or more. When common sense returned in 1637 with the collapse of tulipomania, many investors were ruined. Some jumped into the canals and drowned themselves. But the tulip recovered nicely, becoming one of the most hybridized plants in the history of world botany. The genus has about 100 true species, most native to the Middle East and southeastern Europe. As usual with bulbs, I suggest you try some of the small, exquisite species tulips after you have sated your colour lust on the big, bold Dutch hybrids.

Family: Liliaceae, the lily family

NOVEL LORE

In 1850, Alexander Dumas the elder published a popular bit of French fiction, *La Tulipe Noire*, in which the hero, Cornelius van Baerle, succeeds in growing a black tulip and wins a prize of 100,000 florins. Later hybridizers have produced dozens of tulips of deep-purple hue, which they have called *La Tulipe Noire*. No true black tulip exists. Yet.

HERBS

"They told me it was a magic herb with an effective depilatory action. I was misinformed."

HERBS

WORD LORE & DEFINITION OF HERBS

In Latin, *herba* was any plant not a shrub or a tree. In botany, an herb is any non-woody, seed-bearing plant that dies to the ground after its growing season, and if perennial, sprouts again the following spring. In general English usage, an herb is any plant whose parts are used for curative, aromatic, or cooking purposes. By the way, even the ancient Romans sometimes dropped the *h* and said *erba*, thus producing early the present Italian word for grass, with which compare *l'herbe* with its French meanings of herb, plant, weed, grass, and—most felicitous—picnic lunch, in the phrase *le déjeuner sur l'herbe*, the subject of an Impressionist masterwork by Édouard Manet. Known to all lovers of French cuisine is the *bouquet garni*, a bundle of *fines herbes* tied together, perhaps in a mesh bag, cooked with a dish and removed afterward. *Fines herbes* are also used to decorate and add savour to the surface of a cooked dish.

Herba, the Latin word, made its way into South American Spanish as *yerba*, giving places named after a useful herb like Yerba Buena, California. *Yerba buena* (Spanish, good herb) is a creeper used in southwestern cookery for its minty brio. But it was discovered and introduced to botany at the northern limit of its range in what became southern British Columbia by David Douglas (1798-1834), a Scottish naturalist and collector of botanical specimens. The Douglas Fir is named after him, and so is the yerba buena, as *Satureja douglasii*. Douglas introduced the most number of North American plants to international botany, and some 50 species bear his name. In 1825 he reached Fort Vancouver and then collected along the Columbia, Saskatchewan, and Hayes rivers, travelling to meet Sir John Franklin's expedition on Hudson's Bay in 1827. Two years later he returned to collect along the Okanagan and Fraser rivers.

An Elizabethan herb called Hollow Root. The Latin reads: "Hollow Root of herb gardens or Purple-flowered Birthwort." But it is not a member of the modern Birthwort family.

Consider too *yerba maté* from which a pleasing tea is made, the restorative infusion favoured by Argentinian gauchos. *Hierba* survives in Castilian Spanish.

In this section bloom familiar friends like rosemary, thyme, and sage, culinary herbs that can be grown in many Canadian gardens. One might devote a whole bed to these tangy delights, perhaps the patch of garden nearest the kitchen door. Or, one might wish to get serious and tend a herbarium with dozens and dozens of varieties of angelica, borage, lemon balm, and sweet marjoram. As well as meaning herb garden, herbarium also names a collection of dried specimen plants mounted on sheets of stiff paper.

It is not for this little introduction to the etymology of common herb names to wax rhapsodic about the superiority of freshly picked herbs to those dried and ground and sold in a bottle. If you know the difference in zippiness between freshly ground peppercorns and pepper in a bottle, you don't need to be reminded of the potent zest of fresh rosemary as compared with the flat, wan hint of palely residual rosemary that lingers pathetically in powdered preparations of the herb.

Of course, one can go too far. There are herb and spice snobs, persons who invite you to dine in their homes, and greet you at the door talking like cookbooks: "I know you'll *love* my entrée tonight. It's so diverting to be served chops *en croûte*, trigly strewn with a feathery garnish of chervil leaflets." At such a point, it is best to step back, scratch your head in a bumpkin-like manner, and mumble, "Gee, I usually just shake sand on my chops. You do have sterilized beach sand, don't you?" The true herb snob will top you with: "But, *naturellement*, Bill. I use a soupçon of sand myself, often when preparing harissa, you know, that North African spice mixture for couscous stews. Why, just last week I literally force-fed my dear husband an intriguing little recipe I found stamped on the back of a brick of Tunisian hashish. So simple really. First, obtain a gazelle's vulva. Then, marinate in red wine and cardoon essence for ..." But I didn't hear the rest. I had tiptoed out of their house, suddenly remembering an appointment for root-canal surgery which I simply couldn't pass up.

See *Anchusa* in *Annuals*.

ALKANET

❧

Common names of species with herbal use: Chives, Garlic, Leek, Scallions, Shallot, Top Onion

ALLIUM

Genus: Allium < *allium* Latin, garlic < *all* Celtic adjective, hot, pungent, spicy. From the Latin comes the French word for garlic, *ail*, and Italian for garlic, *aglio*. This large genus contains more than 300 species. For ornamental onions like moly and giant allium, see *Allium* entry in *Bulbs* section.

Family: Alliaceae, the capacious onion family

SPECIES
Shallot
Allium ascalonicum is one horticultural designation for the shallot, the fine little onions of French cookery, mild in flavour and aroma, used in salad dressings, chopped and sprinkled on steak, and with dozens of uses in *haute cuisine*. The shallot has a compound bulb, with each "clove" wrapped in a papery, purple tunic. The specific is Botanical Latin meaning 'of Ashkelon,' which was one of the five cities of the Philistines, a harbourless seaport in the land of Canaan. The other cities were Ashdod, Ekron, Gath, and Gaza. Known as Askalon and Escalone in European languages, it was the site of a victory by the Crusaders over the Egyptians in A.D. 1099. It is mentioned in one of the most quoted phrases from the Old Testament, in the second book of Samuel 1:19-20, when King David laments the deaths of Saul and Jonathon: "How are the mighty fallen! Tell it not in Gath, publish it not in the streets of Ashkelon; lest the daughters of the Philistines rejoice."

WORD LORE OF SHALLOT
Shallot < *échalotte* Early Modern French < *eschalotte*, *eschalette* Medieval French < *eschaloigne* Old French = *scalogno* Italian < **scalonia* Late Latin < *caepa ascalonia* Latin, onion of Ashkelon.

Note that from *escalogne*, an Old Norman French form of this common onion word, Middle English derived *scalyon*, which in turns gives another onion word still used in English, *scallion*.

Adding a Bit of Confusion

The perennial forms of the common onion, *Allium cepa* (*caepa* Latin, onion) are sometimes called shallots. But these have no connection with Ashkelon, nor with the delicate mildness of taste of true shallots. The true botanical name of shallots is partially responsible for the present confusion, because that true name of shallots is *Allium cepa aggregatum* (Latin, clustered, said of the small bulbs).

Cocktail Onions, Egyptian Onions, Top Onions

Allium cepa proliferum (Botanical Latin < *prolifique* Scientific French coinage, producing many offspring < *proles* Latin, children, progeny) is the top onion, an oddity in appearance, with a crown of small, brown-tunicked bulbils at the top of a stem sometimes a metre high. As the weight of this bulb-clustered head increases, the stem bends down, touches soil, and the individual little bulbs take root.

Leek

Allium porrum (Latin, leek) is the biennial onion whose outer surface collects sand and soil grit. Always wash leeks well. The Latin *porrum* is also the source of the French word for leek, *poireau. Porrum* has the basic meaning of sword-poke or spear-poke, in reference to the shape of the leaves. The Latin word for leek is akin to the ancient Greek word for leek, *prason*, and possibly to a group of verbs in Germanic and Scandinavian languages such as *porren* Dutch, to poke as with a sword, and Danish *purre*, to prod, to thrust. A porr was once a fire poker in English, and we had a verb too, to porr = to poke something with a spear. The habit of growth of members of the onion family has seen them named after their long, sword-like leaves, for example, spear-leek in Old English was *gar-leac*, which has the current spelling: garlic.

The leek is the plant symbol of Wales, and an essential of French cookery. The somewhat soft bulb and the lower leaves of leeks are used to flavour soups, stews, most versions of vichyssoise, and served by themselves cooked with lemon butter. Leeks are also boiled, chilled, and presented *en vinaigrette*. Legend says the Welsh symbol arose during a battle in which Cadwaladr, King of Wales, ordered his soldiers to stick sprigs of leek in their caps, so that during the tumult on the battlefield Welsh troops could be distinguished from those of their enemy, led by King Edwin of Northumbria. The Welsh did so and won the engagement, and the symbolic leek recalls this victory.

WORD LORE OF LEEK

Leek < *leac* Old English, leek, akin to *locc* Old English, a lock of hair, itself related to *lokkr* Old Scandinavian, hank of hair, and to *lugos* Greek, a pliant twig, a bendable stick of wood.

Leek, Allium porrum

Garlic

Allium sativum (*sativus* Latin, cultivated, literally 'able to be sown or planted' < *serere* Latin, to sow) is common garlic < *garleac* Old English, spear-leek. Compare the fish with the long, spear-like snout, the gar or garfish.

One of the earliest mentions of the curative properties of garlic is in the Ebers papyrus from about 1500 B.C. where cancer is described and one of the treatments for hardened skin cancers is the external application of garlic paste. Hippocrates prescribed eating garlic as a treatment for uterine cancers. Some modern cancer research, still inconclusive, does involve garlic and other members of the onion family. And yes, your Polish grandmother was correct, garlic really is good for you. Among the chemical goodies in those little cloves are antibiotics like allicin, a powerful bactericide, and allistatin, a broad-spectrum fungicide.

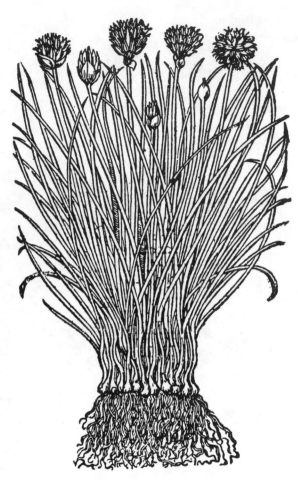

Chives

Chives

Allium schoenoprasum (Botanical Latin < *schoinos* Greek, rush, reed + *prason* Greek, leek) is chives, one of the smaller onions, whose fresh leaves have been chopped and added to various foods for more than five thousand years. As one of the *fines herbes* of French cuisine, chives give a modest hint of onion to many a sauce or main dish. Chives perk up sour cream, cottage cheese, cream cheese, and a moderate sprinkle will brighten an omelette.

The word *chive* entered English from an Old French dialectal variant of Old French *cive*, itself from *caepa* or *cepa* Latin, onion. Compare *chivot*, the word for green onion in the northern French dialect of Picardy.

Scallions

Green onions, spring onions, scallions are really just immature seedling onions, plucked before they mature, to preserve the tender hint of onion they deliver to chopped salads, herb butters, and dozens of other dishes. Scallion derives from the same Latin word as shallot. Shallot < *échalotte* Early Modern French < *eschalotte*, *eschalette* Medieval French < *eschaloigne* Old French = *scalogno* Italian < **scalonia* Late Latin < *caepa ascalonia* Latin, onion of Ashkelon. Then note that from *escalogne*, an Old Norman French form of this common onion word, Middle English derived *scalyon*, which in turns gives *scallion*.

❧

ANGELICA

Common names in British dialects of the wild plant: Angel's Food, Kewsies (Lincolnshire), Spoots (Shetland), Water Squirt (Somerset)

Genus: Angelica < *herba angelica* Medieval Latin, angel's plant. Widespread across Europe and Asia, angelica was an important cure-all of the herbalist, hence its angelic name. A triennial, angelica blooms in its third year, then sets seed and expires, joining the Choir Botanical.

Family: Umbelliferae, the carrot family, named from the flowering habit of its members, in an umbel, that is, an inflorescence in which all the little flower stalks arise from a central point and form a rounded cluster, as in the flower of our common ditch weed, Queen Anne's Lace. Several members of this family are important foods, like carrot, celery, and parsnip. Others are herbs and spices like anise, caraway, coriander, dill, fennel, and parsley.

SPECIES

Angelica archangelica (Latin, of the archangel) is a giant among herbs, often reaching one or two metres high in cultivation. The specific recalls an old myth that the medicinal virtues of the plant were revealed to a monk collecting simples (herbs which could be used all by themselves) by the archangel Raphael as a sure cure for the plague. It is not; but angelica was held in such high esteem that its synonym in medieval Latin was *radix Spiritus Sancti* 'root of the Holy Ghost.' The thick, hollow stalks are candied and make a peculiarly sharp-tasting treat all across Europe. Tasting somewhat like juniper berries, angelica seed oil and root extract help to flavour some gins, liqueurs, and vermouths, and are added to perfumes to counteract the cloying, honeyed sweetness of other aromatics. The bitter leaves are in some areas boiled and eaten like spinach. One modern use of note occurs in Britain and Europe, where some drivers pin up a mesh bag of crushed leaves in the interior of automobiles and swear that angelica reduces car sickness. Perhaps it is effective against less common forms of travel nausea, such as the one I experience whenever I find myself within one kilometre of a theme park?

Diaghilev & Angelica

So high was medieval demand for angelica, to protect against plague, witches, and poison, as well as to stir passion as an aphrodisiac, that the plant name was used in some languages to form an occupational surname. The most famous person to bear such a last name was the Russian ballet impresario, Sergei Diaghilev, founder of the Ballet Russe company in 1909, introducer of Nijinsky to Paris, commissioner of music from Stravinsky. His surname means 'descendant of a man nicknamed Angelica,' probably because the ancestor of the family sold herbs. The Russian word for angelica is *dyagil*.

∾

ANISE

Other common name: Aniseed

Genus: Pimpinella, fourteenth century herbalist's Latin < *pimpernelle* Old French < *piprenelle* Proto-Romance < *piperinus* Late Latin, a diminutive form of *piper* Latin, pepper. The seed capsule of some plants called by these names resembled a peppercorn. Anise seeds or aniseed smell and taste like licorice, and have been used for millennia, chewed as breath fresheners and put as flavouring into pastries, candies, liqueurs, and cordials.

Pimpinella also gives a name to a plant of a different family, the Scarlet Pimpernel, a flower chosen as his symbol by the dashing hero of a turn-of-the-century potboiler by Baroness Orczy, *The Scarlet Pimpernel* (1905), in which the elusive title character saved French aristocrats from the guillotines of the Revolution by smuggling them out of France.

The verbal root is among the oldest spice terms in the Indo-European language family, appearing in ancient pepper words like Greek *peperi* and Sanskrit *pippali*.

Family: Umbelliferae, the carrot family

SPECIES

Pimpinella anisum or common anise is a challenge to grow in a Canadian garden, because the seeds need a long summer (four months) to fully ripen and because the delicate roots are very fussy about being transplanted. But with due diligence Canucks can enjoy the plant as a sweet-smelling herb. The specific is Latin and has a variant *anethum*. See *Dill* entry in this section. Both were borrowed from Greek, *anison* and *anethon*. All the terms were applied to herbs of the *Pimpinella* genus but also to a variety of plants from other families.

Uses

Whole or crushed aniseed zests up desserts, curries, and pickles. A medieval comfit was made by coating the seeds with layer upon layer of crystallized fruit sugar. The flowers and young leaves are added sparingly to fruit salads and soups, and they are used to garnish vegetables. The most common commercial use of aniseed is to flavour alcoholic drinks, such as anisette, pastis, Pernod, and Ricard in France, ouzo in Greece, raki in Turkey, and arrak in Asia and the Middle East < *'araq* Arabic, sweet juice, liquor. On the home bar, some adventurers put a teaspoon of anise seeds into a pint of brandy, and let them steep for several weeks, to add a licorice flavour. Remember however that in large doses anise oil is toxic.

❧

Other common name: Sweet Basil

BASIL

Genus: Ocimum, Latin < *okimon* Greek, an aromatic herb, perhaps our basil, perhaps not

Family: Labiatae, the mint family < *labiatus* Botanical Latin, lipped, with a prominent lip < *labia* Latin, the lips. In botany, the labium is the lower lip of a flower with two lips. In flowers of the mint family, this labium is highly developed and enlarged, while the upper lip of the corolla though present is rudimentary. The mint family is large, with 224 genera and 5,600 species in the tropics and warmer temperate zones.

SPECIES

Ocimum basilicum is Sweet Basil. Its most common name in ancient Greece was *basilikon* 'the king's herb, the royal herb' < *basileus* Greek, king. A *basilike* was first a Greek royal palace, then the word was borrowed into Latin, and basilica took on religious meanings. Sweet Basil was perhaps used in some ancient medicine, in earliest days made only by a king, or perhaps the important plant, even wild, was considered the king's property, and could be collected only by regally ordained agents. The introduction of tomatoes to Europe produced a revival in the popularity of basil, when continental cooks discovered the succulent symbiosis of the fruit and the herb.

Pesto

A gift of Genoese kitchens to the world is pesto sauce, in which basil leaves are pounded, hence the name, from *pestato* Italian, pounded < *pestare* to crush, to grind. For a related word, think of a mortar and pestle, ultimately from Latin *pistillum* 'little pounder.' One pesto is made by pounding into a green, delicious paste basil leaves, Parmesan cheese, garlic, olive oil, and pine nuts.

❧

BAY LEAF

Other common names: Bay, Bay Laurel, Sweet Bay

Genus: Laurus < *laurus* Latin, the Bay Laurel tree. This appears to be an ancient Mediterranean tree word. Attempts have been made to connect it with *laura* Greek, lane, pathway,

Bay leaf or Laurus

as the tree that grows along lanes, and with the Greek place name *Laureion* or Mount Laurium, a promontory south of Athens famous for its silver mines. What we do know is that the laurel was sacred to Apollo, and thus prize-winning Greek and Latin poets and soldiers wore a wreath made of intertwined laurel sprigs. Winning athletes in classic contests like the Pythian games donned crowns of bay leaves. Such a victor was dubbed in Latin *laureatus*, which gives our English phrase, poet laureate.

The personal names Laura and Lawrence both derive from *laurus*, Lawrence finding its start in the name of a saint who came from an area of the Roman empire noted for its bay trees, Laurentium. Canadian Prime Minister Wilfrid Laurier had a surname derived from an ancestor who lived beside a stand of bay trees. The Greek word for bay tree was *daphne*. In Greek mythology, Daphne was a comely nymph pursued by Apollo. She escaped rape when she was transformed into a bay tree.

Family: Lauraceae, the bay tree family

SPECIES
Laurus nobilis (Latin, noble, stately, notable) is a large tree in native Mediterranean soil. In North America it is usually kept as a pot shrub that spends winter indoors. Its sweet bay leaves can be cut fresh, used fresh, or dried. Bay leaf is one of the herbs in a *bouquet garni*. Marinades, pâtés, soups, and stews boast bay as well.

WORD LORE
In early English, bay referred to the berry or the tree of laurel < *baie* Old French, the berry of the bay tree < *baca, bacca* Latin, berry, bay berry < ? *Bacchus*, in whose festivals 'the king of the cups' was crowned with laurel.

∽

See *Bee Balm* entry in *Perennials*.

∽

BERGAMOT, BEE BALM

BORAGE

Other common names: Bee Bread, Burrage (British dialect)

Genus: Borago, Botanical Latin < *borrago* Medieval Latin < *abu 'araq* Arabic, father of sweat, of moisture, of liquid—because early Arabic medicine used borage as a sudorific or diaphoretic, an agent that causes sweating. The English version *borage* < *bourrache* Old French < *borrago* Medieval Latin.

The mature plant is covered with stiff, prickly hairs. Consequently only the young leaves that taste pleasantly of cucumber are used to flavour drinks and salads. These rough, woolly hairs may account for a second, possible origin of the medieval Latin word, namely in the Old French *bure*, *bourre*, and *bourre de laine* 'homespun cloth of brown wool.' It was the prickly material from which poor monks' robes were made, the same monks who would have collected borage in the wild and grown it in their herb gardens.

A long prominent Québec family takes its name from this cloth. One who made and sold *bourre de laine* had the French occupational surname of *bourrassier*. Bourassa is a regional variant, and the surname of Henri Bourassa (1868-1952), founder in 1910 of *Le Devoir*, one of Canada's most influential newspapers. Henri Bourassa was an important Québec politician who advanced French-Canadian nationalism. He was the grandson of Papineau. Robert Bourassa was Premier of Québec (1970-1976 and 1985-1994) and helped draft the Meech Lake Accord and supported the Free Trade Agreement.

Bourre or *bure* came into Old French from the popular street Latin of the Romans who conquered ancient Gaul. There *burra* meant 'rough wool' or 'a shaggy garment.' *Burrus* was an old Latin adjective for brownish-red, the colour of such clothing. Incidentally, it was also the colour of a small donkey used as pack animal, and Roman soldiers posted to the Iberian peninsula called the animal *burrus*, thus planting the verbal seed for one of the earliest words in the Spanish language, *burro*. In Mexico, a tortilla wrapped around a filling of spiced beef and other yummies looked to eaters like a little donkey loaded down with a colourful pack, and so the diminutive form *burrito* meaning 'little donkey' came to be applied to the food as well. *Burrus* was borrowed from or was akin to the Greek colour adjective *pyrros* 'fiery red' whose root is *pyr*, which is cognate with English 'fire.' Compare these English words: pyre, Pyrex™, and pyromaniac.

Family: Boraginaceae, the borage family, a small group of European and Asian herbs grown for red dye from their roots and for their blue flowers, which are still used to garnish salads. Borage leaves are used to flavour punches and cocktails like Pimm's Cup.

Borago semper virens.
Neuer dying Borage.

SPECIES

Borago officinalis is the annual herb with drooping blue, white, or pink flowers, often therefore planted up on a slope, so the starry flowers are more easily seen. Borage is a prolific self-seeder, and will appear year after year in the same patch. *Officinalis* is an interest-ing medieval Latin adjective seen as the specific name of many older botanical plants that were sold in shops because they had medicinal and cosmetic uses. *Officina* is the medieval Latin word for a workshop or shop, from which English gets of course *office*. *Officina* is a contraction of *opificina* from *opifex* 'craftsman, mechanic' which in turn is made up of *opus*, Latin, work + *fex*, *ficis* Latin noun and adj. suffix, doing, making. Compare *facere* Latin, to do, to make. Officinalis as a specific in a plant name almost always indicates the plant was widely used in medieval times and often centuries before for some human purpose. Borage was often added as a flavouring to tankards of wine and cider. The ancients believed borage essence drove away melancholy and gladdened the heart. Medieval students had the notion that a few young leaves of borage in their dinner wine would cheer them on to further study.

❧

CATNIP

Other common name: Catmint (British)

Genus: Nepeta < *nepeta*, Latin word for Italian catnip < *Nepeta* Etruscan, an ancient city in Etruria, modern Tuscany, that may have supplied herbs to Rome, or invented some use of the plant.

Family: Labiatae, the mint family < *labiatus* Botanical Latin, lipped, with a prominent lip < *labia* Latin, the lips. In botany, the labium is the lower lip of a flower with two lips. In flowers of the mint family, this labium is highly developed and enlarged, while the upper lip of the corolla though present is rudimentary. The mint family is large, with 224 genera and 5,600 species in the tropics and warmer temperate zones.

SPECIES

Nepeta cataria takes its specific from the medieval herbalist's name for this plant, *herba catti* or *herba cattaria*. Note that Linnaeus misspelled the specific, using only one *t*, but botany stubbornly insists, as always, on keeping such mistakes in its official nomenclature. The English word *catnip* does not arise from cats nipping at the plant (which they do with tireless gusto) but from a shortening of *nepeta* to *nep* and then to *nip*. In the days when Britons practised sadistic corporal punishment, catnip also meant a lash or forty with a whip called a cat-o'-nine-tails. Cats go crazy about catnip because the leaves and the root contain a chemical closely related in structure to certain feline sexual pheromones, odorous organic chemicals that animals give off to stimulate behavioural responses from other members of their species, responses that include sexual arousal. Manufacturers of toys for cats often drench their products with catnip extract. A refreshing mint tea for humans is made from the young leaves. But fret not. The pheromones are species-specific. One can drink such tea safely, in the assurance that one will not be expelled from the neighbourhood for interfering in an untoward manner with family pets.

❧

CHAMOMILE

Other common spelling: Camomile

Genus: There are two generic names. The older word, but the newer botanical genus name is Anthemis < *anthemis* Greek, chamomile < *anthemon* Greek dialectal variant, flower < *anthos* Greek, flower. From its many uses in Greek herbal medicine, both leaves and flowers, it was simply *the* flower.

The second genus name is Chamaemelum < *chamaimelon* Greek, chamomile < *chamai* Greek, on the ground + *melon* Greek, apple. It was called ground apple because the flowers do have an apple-like aroma and because of the low-growing, mat-forming habit of growth of many of the 100 species.

Family: Compositae, the daisy family < *compositus* Latin, placed together. This largest family of flowering plants is named after the compound flowers of its members. Small florets of individual flowers make up large clusters or heads.

SPECIES

Anthemis nobilis or *Chamaemelum nobile* (*nobilis* Latin, notable, famous, renowned, excellent) has been used for thousands of years in the form of a soothing, sedative tea made from the dried flowers.

∽

Genus: Anthriscus, Latin < *anthriskos* Greek, literally 'little flower' < *anthos* flower + *-iskos* common diminutive suffix, little. Several chervil-like plants had variants of this word as names in ancient Greek, for example, *anthryskon* and *anthriskion*.

CHERVIL

Family: Umbelliferae, the carrot family

SPECIES
A Botanical Name That Is Wrong, Wrong, Wrong
Chervil is *Anthriscus cerefolium*, sometimes seen as *cereifolium*, both examples of an erroneous species name, sanctified by long usage, which botanists are unwilling to change. There are several errors connected with the scientific names of chervil, and, yes, I'm going to have to wag the nitpicker's finger here again. Now *cerefolium* is a specific adjective in Botanical Latin and it does mean 'with waxy leaves.' But it is wrong here for two reasons. First, there is nothing whatsoever waxen about the soft, crisp, parsley-like leaves of chervil.

Second, *cerefolium* is an old mistake by someone who knew a little Latin and very little Greek. *Cerefolium* is a mistranslation of the old Greek name for chervil, *chairephyllon*. The copyist thought it looked like *cerum* Latin, wax + *folium* Latin, leaf. No. *Phyllon* is the Greek word for leaf, or *folium* in Latin. But the basic meaning of the common Greek verb of greeting, *chairein*, is: take pleasure in, rejoice, be healthy, thus *chairo-phyllon* 'healthy-leaf' or 'leaf of rejoicing.' Both are apt concepts for the name of an herb with many medicinal uses among the ancient Greeks. Our English word *chervil* comes from the same Greek name: chervil < *cerfille* Old English < *caerefolium* Late Latin < *chaerephyllum* Latin < *chairephyllon* Greek.

Chervil or Anthriscus

Uses

Chervil is one of the four herbs in the traditional *fines herbes* of French cookery. The other three are chives, parsley, and tarragon. Chervil is interesting to add to certain soups where it blends well with other herbs. Experts say the most savoury leaves are young ones from plants grown in light shade.

CORIANDER & CILANTRO

Other common names: Chinese Parsley. Coriander has broad, lobed lower leaves that are sold as cilantro, to flavour or garnish foods. The upper leaves, seeds, and roots are coriander.

Genus: Coriandrum, Latin < *koriandron, koriannon* Greek, this plant. Some experts like William T. Stearn say this Greek plant name contains *koris* Greek, bug, "referring to the unpleasant smell of the unripe fruits which disappears when they are ripe and dry." I don't agree. First, *koris* was not just any bug, it was a bedbug. There is no reference in ancient Greek literature to bedbugs having a bad smell. Second, Professor Stearn blithely ignores *-andron*, the other verbal root in this compound plant name. Third, *koriandron* probably represents a word borrowed into Greek from some language of the Middle East, perhaps Egyptian, but glossed by the Greeks as if the word were composed of *kore* Greek, maiden + *aner*, *andros* Greek, man. The Greeks called this plant 'maiden and man' because it has two different kinds of leaves on the same plant. The upper leaves are finely cut and parsley-like (to the Greeks, maidenly); the lower leaves, sold as cilantro, are broad, stubby, and only slightly incised (to the Greeks, manly).

Professor Stearn is the world's leading authority on Botanical Latin, so it is always disconcerting to observe the mistakes in Greek etymology present in his books. Not to belabour the point, but, in his *Dictionary of Plant Names for Gardeners* (1992), just across the page from his derivation of coriander from *koris*, is his origin of the name of the cosmos flower. Stearn says it is from "GR. *kosmos*, beautiful." Is it indeed? *Kosmos* is a noun in classical Greek, not an adjective, and it means 'good order, ornament, good arrangement.' *Kosmos* does not mean 'beauty.' There is an adjective from the noun *kosmos*. It is *kosmios* and it means 'decorous, well-behaved, in regular order.' Some of the Greek adjectives for beautiful were *kalos*, *eumorphos* (well-formed), and *kallimorphos* (beautifully shaped). The noun *beauty* was usually rendered in classical Greek as *to kallos*. Has Professor Stearn ever condescended to check his suppositions in an authoritative dictionary of classical Greek, and to follow them up by reading the classical passages citing the Greek words? Or, are his readers to accept his unreferenced notions as they drop from his pen? For the correct derivation, see *Cosmos* entry in *Annuals*.

Family: Umbelliferae, the carrot family

SPECIES

Coriandrum sativum (*sativus* Latin, cultivated, literally 'able to be sown or planted' < *serere* Latin, to sow) has been used as a spice for more than ten thousand years, extensively over its broad, native range, which runs from North Africa through to western Asia. Egyptian papyri mention coriander root as an aphrodisiac. Cooks in India made fresh coriander root a key ingredient in curries, and extracted seed oil to make a spicy incense. In the Old Testament, Exodus 16:31, one reads: "And the house of Israel called the name thereof Manna: and it was like coriander seed, white." The ancient Greeks flavoured wine with coriander leaves and seeds. Coriander is one of the flavouring agents in some modern toothpastes.

WORD LORE OF CILANTRO

The letter *r* sometimes shifts in dialectal borrowings to *l*. This slightly unstable *r* is common when certain words enter one language from another. Thus, the classical Latin *coriandrum* had, by the time of medieval Latin, a *r:l* shift form, *celiandrum*, which produced *cellendre* in Old English and *cilantro* in Spanish, from which English borrowed the word to refer to the lower leaves of the plant used as an herb. English reborrowed the *r* form from *coriandre* in Old French, to give the modern English *coriander.*

DILL

Genus: Anethum < *anethum* Classical Latin, the word for dill or anise in Vergil and Pliny < *anethon* Greek, dill < possibly *anienai* Greek, to remit (of a disease), hence the English medical term borrowed from Greek, anesis, remission of the symptoms of a disease. *Anetikos* was an adjective applied to herbs that meant 'soothing, lessening the severity of a disease.' The plant may have been so named because of its many uses in ancient Greek herbal medicine. Dill had a synonym in older English, anet.

Family: Umbelliferae, the carrot family

SPECIES

Anethum graveolens (Latin, strongly odorous) is dill, a tender annual whose faintly sweet leaves and seeds were used in ancient times as a carminative, an agent that encouraged farting, to help expel intestinal gas built up in colic or gripe. Dill leaves,

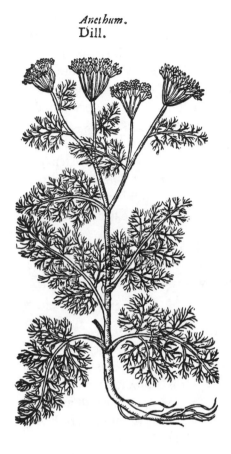

Anethum.
Dill.

best when fresh, still flavour soups, vegetables, salads, fish, and sauces. The slightly bitter seeds are a principal pickling spice, and are also used in sauerkraut and some vinegars.

WORD LORE OF DILL
Dill is the Proto-Germanic name for the plant, its oldest reflex appearing in Old High German *tilli*.

❧

DULSE

Dulse is an edible seaweed, a red alga, not technically an herb, but it belongs in any Canadian list of yummy plants.

Other common name: Salt Leaf

Genus: Palmaria, Botanical Latin < *palmaris* Latin, the width of a hand, also influenced by *palmatus* Latin, with lobes that resemble a human hand with the fingers stretched out. Dulse has dark red, palmately divided fronds.

Family: Rhodophyta, the red algae family < *rhodo-* Greek word element, rose-coloured, red < *rhodon* Greek, a rose + *phyton* Greek, plant

Rhodophytes

The Canadian coast has about 175 arctic, 350 Atlantic, and 500 Pacific species of seaweed, many of them algal. Algae are divided into greens, browns, and reds. Most seaweeds are red and brown algae. Rhodophytes are a division of the algae, which store floridean starch, a compound like amylopectin, that is, branched chains of polysaccharides, glucose units joined by glycoside bonds. This makes them sweet, mucilaginous, and sticky. Many commercial products of a glue-like consistency are made from red seaweeds, for example agar-agar, algin, and carragheen. They are hydrocolloids used to thicken, stabilize, or emulsify products like ice cream and toothpaste.

SPECIES

Dulse is *Palmaria palmata*, which grows on both shores of the North Atlantic Ocean. Longfellow wrote that "the tide is low, and the purple dulse is lovely."

WORD LORE OF DULSE

Dulse was borrowed directly into English at a rather late date,circa 1698, from Scots Gaelic *duileasg* < *duil* Gaelic, leaf + *iasc* Gaelic, fish, hence in origin it is a playful, Celtic nick-name, leaf-fish, because it came from the sea like fish and was an edible leaf. In Welsh, it's *delysg* or *dylusg*, in modern Irish (Erse) *duileasc*. Sometimes in Scotland, it's spelled *dilse*.

Habitat

Along the coasts of our Canadian Maritimes, dulse grows in the intertidal zone at the low-water mark, attached to rocks or other seaweeds by a holdfast shaped like a disc. Many Canadians say the best dulse in the world is harvested along the rocky shore-line near Dark Harbour, on the west side of Grand Manan Island, New Brunswick.

Uses

Although it is called tough and flavourless by those who dislike it, dried crisps of dulse are chewed by themselves and added to soups and stews as salty thickeners. The word gave rise to an occupational noun in English, e.g., "The dulseman wheeled his slimy boxes to the top of the brae." In England, all edible sea-weeds are sometimes called *laver*. Welsh laverbread is made by boiling dulse or other seaweeds, mixing it with oatmeal, and frying it in bacon fat or butter.

∾

FENNEL

Genus: Foeniculum, Botanical Latin < *faeniculum* Latin, fennel < *faenum* Latin, hay + *iculum* Latin, diminutive ending, mean-ing 'little,' 'little hay' in reference to the fancied resemblance of fennel to hay or because it grew as a weed in hay fields. The Indo-European root in the Latin word for hay, *faenum*, is **fe* 'making offspring' which appears in other words like fecund, feminine, and fetus. Much reduced, *faeniculum* entered Old French as *fenoil*, then was borrowed into Old English as *fenol*, eventually to produce the modern English spelling.

Fœniculum vulgare.
Common Fennell.

Family: Umbelliferac, the carrot or parsley family

SPECIES

Foeniculum vulgare (Latin, common) is a tender perennial commonly reaching a metre or more in height. Its feathery, green, faintly licorice-scented leaves and its seeds are among the oldest known culinary herbs, used particularly to flavour fish. In Italy, a related but smaller species, *Foeniculum azoricum* (Botanical Latin, of the Azores islands), Florentine fennel, is used as a vegetable. Its bulbous leaf base is used raw in salads or cooked as an indi-vidual vegetable serving. Fennel seeds are used in baked goods, sauces, and to flavour some European liqueurs. Italian hot and sweet sausages, the freshly made kind, often sing with fennel seed.

The Italian for fennel is *finocchio*, which is also the common and insulting slang term for a homosex-ual male in Italy, making it the equivalent of English putdowns like fag, fairy, and queer. It originally denoted the stereotype of the homosexual male who was overdressed or cross-dressed, all feathery and plumaceous like fennel leaves.

∾

LAVENDER

Other common names: French Lavender, Spike, Spike Lavender, Stickadove

Genus: Lavandula, Medieval Latin, any of several lavender species. The generic has two possible origins. *Lavandula* may be a diminutive based on a gerundive noun of the Latin verb *lavare* 'to wash' where *lavanda* would mean 'things to be washed, things pertaining to washing.' Lavender has been used since ancient Egyptian times to perfume toiletries and to scent washables. For more than two thousand years, the essential oil has been added to bath water. Lavender did enter English through the Anglo-Norman French form *lavendre*.

Or, *lavandula* may stem from the lilac-purple colour of the flowers in many species. One of the Latin adjectives for this colour was *lividus*. A recorded, medieval Latin, diminutive form of *lividus* was *livendula* 'little purple flower.'

The verb *lavare* gives many current words in English like lather, latrine, laundry, lavatory, and lavish. Here too perhaps is volcanic lava, originally a word in the Italian dialect of Naples applied to any heavy rain that washed through the streets, and then transferred to name what flowed out of Mount Vesuvius. Compound *lavare* verbs end in *-luere* in Latin, and give English words such as ablution, alluvial, antediluvian, and dilute.

Family: Labiatae, the mint family < *labiatus* Botanical Latin, lipped, with a prominent lip < *labia* Latin, the lips. In botany, the labium is the lower lip of a flower with two lips. In flowers of the mint family, this labium is highly developed and enlarged.

SPECIES

Lavandula angustifolia (Botanical Latin, narrow-leaved) is the common lavender native to western Mediterranean lands. Most cultivated lavenders have this as a distant relative.

Lavandula stoechas is sometimes called French Lavender or Stickadove. *Stoechas* was often pronounced 'stickas.' But Stickadove arose as a common name because of the large, stiff, coloured bracts that stick up from the top of each flowerhead on this lavender, not, one hopes, because intemperate persons sought to impale wee dovelets, even when their dolorous cooing disturbed the peace of early morning. The specific indicates one of this plant's native habitats on a little group of islands called the Stoechades, which lie off the south coast of France near Toulon and are called in French *Les Isles d'Hyères*.

Uses

All above-ground parts of lavender have glands that secrete an aromatic oil, with the majority of these glands on the flowers and stems. Lavender has been used for millennia as a sachet, often the whole plant being dried and placed between individual sheets to keep linen fresh. Lavender has flavoured candy, cream, jams, stews, and vinegar. It scents many a bowl of potpourri, and makes a long-lasting, sweet-smelling patch in the perennial garden. Still used in baths to relax the bather with its pleasant aroma is lavender oil. Lavender is mentioned by name as one of the plants taken to America on *The Mayflower*.

LEMON BALM

Genus: Melissa < *melissa* Greek, honeybee < *meli* Greek, honey; bees are attracted to the nectar of lemon balm flowers. The Greek word is akin to *mel* Latin, honey, and to the name of a drink made from fermented honey, mead. Related word forms are widespread throughout the ancient Near East. Compare the Semitic triliteral *mtq* which gives *mathoq* Hebrew, sweet; *matqa* Arabic, sweetness; *matqu* Akkadian, sweet; and perhaps related to *retchem-t* Ancient Egyptian, anything sweet. A mellifluous voice is a sweet-sounding one, literally one that flows like honey. For the interesting etymology of balm, see *Fir* in the *Trees* section.

Family: Labiatae, the mint family

SPECIES

Melissa officinalis (Medieval Latin, of the herbalist's shop, *officina*) is the mint-like, spreading perennial grown for its lemon-scented leaves, which are used to season veal and poultry, to brew a refreshing tea, and whose essential oil is one of the constituents of commercial perfumes. Lemon balm is sometimes confused with lemon verbena, which is *Lippia citriodora*, a shrub difficult to grow in Canadian gardens, whose specific epithet means 'lemon-scented.' Lemon mint is a different plant too, *Monarda citriodora*. See *Bee Balm* in *Perennials*.

LOVAGE

Other common name: Love Parlsey

Genus: Levisticum, Botanical Latin < ultimate origin unknown, although, with no proof, some authorities say it is a corruption of the name of a similar plant, Ligusticum, literally 'plant from Liguria in northwest Italy.' The problem? There is not a jot or tittle of printed proof. We do know the following trail: *levisticum* Late Latin > *levesche*, *luvesche* Old French > *lovache* Middle English, with the old French form glossed by English ears as if it contained *ache*, an obsolete English word for parsley, so that the English heard it as a word meaning 'love parsley,' still one of lovage's common names in some English dialects. Modern French is *livèche*.

Family: Umbelliferae, the carrot or parsley family

SPECIES

Levisticum officinale (Medieval Latin, of the herbalist's shop, *officina*) is the giant of herbaceous perennials, reaching mature heights of two metres after four years of good growth. Lovage is vaguely celery-like in form, taste, and use. The dark, green leaves are salad-bound from many an herb garden. Soups and stews perk up with a judicious leaf or

Leuisticum vulgare.
Common Louage.

two. And lovage seeds add a celery flavour to many dishes. All parts of lovage are harmful to pregnant women.

Before I delved into the world of garden herbs and when I first heard the word *lovage* and did not know its meaning, I guessed at it. Lovage sounded like baggage or tonnage, so I supposed that lovage was the emotional price paid by those who love, the burdensome weight of caring deeply about others. The word has a cumbrous solidity of sound. But that was long ago, before I had discovered the joy of loving that easily counterbalances the emotional C.O.D. required.

Genus: Mentha < *mentha* Latin, mint < or akin to *mintha, minthe* Greek, a mint plant; hence the chemical names like menthol and menthane

Family: Labiatae, the mint family < *labiatus* Botanical Latin, lipped, with a prominent lip < *labia* Latin, the lips. In botany, the labium is the lower lip of a flower with two lips. In flowers of the mint family, this labium is highly developed and enlarged, while the upper lip of the corolla though present is rudimentary. The mint family is large, with 224 genera and 5,600 species in the tropics and warmer temperate zones.

THE MINTS

SPECIES

All the mints are easy garden subjects, and all are invasive, spreading rapidly by wandering, deep roots. So it is best to contain mints from initial planting, either in a separate bed or in containers. And don't fertilize mint with manure, which may harbour a mint-rust fungus that smuts up the leaves you want to use.

I recall vividly my first taste of mint tea, the world's most restorative infusion. I had trod for hours on large, blistering cobblestones, under a desert sun that had the thermometer boiling at 43° C. through the main public square of Marrakech in southern Morocco. It is called *Djemaa el Fna*, Arabic for 'meeting place of the dead,' named because early potentates used to display the severed heads of those who had displeased them on sharp stakes that once ringed the square. Nowadays tourists meet there too, in the teeming carnival of Moroccan life, often late in the afternoon, after visiting the sites and after being sunned to a frazzle. Don't be taken in by those who say, "Yes, but it's a dry desert heat." We plunked exhausted at a small café in the square. A djellaba-clad waiter suggested in Moroccan French that we all should *boire une tasse de thé à la menthe*. We watched him make this mint tea, jamming a brass teapot cramfull of fresh spearmint leaves, then pouring scalding hot water over the leaves. After steeping for five minutes, he dropped a peculiar, pyramid-shaped, miniature sugar loaf into the chartreuse brew and, holding a perforated wooden filter over the pot, poured each of us a vaporous cup. Breezy, minty bliss soon sent scurrying any thoughts of heat exhaustion. After several more cups, we were perked up anew and ready for a long, enjoyable evening at the Moroccan National Folk Festival. I still take mint tea on a hot August afternoon, and it is the best use of this herb I know, certainly superior to the use of hideous mint jelly gobbed over a chop to disguise elderly mutton. Of the dozens of species of mints, I list only a few.

Mentha piperita officinalis is peppermint, so often crossed with other mints that it is sterile and therefore usually grown from stem cuttings or root divisions. The first printed reference to peppermint candies is late in the seventeenth century.

Mentha rotundifolia (Botanical Latin, with rounded leaves) is the delightful apple mint. A tall plant with wrinkled leaf surfaces covered with dense white hairs, its flowers and leaves are pleasantly aromatic in cut flower arrangements.

Mentha spicata (Latin, with a spike, said of the flowerhead) is the most widely used of all the mints, spearmint, originally spire-mint, also descriptive of the flowerhead. Spearmint is the classic ingredient in mint juleps, which may actually be consumed in locations other than southern verandahs dripping with bougainvillea and hibiscus. Spearmint leaves brisk up salads, add brio to vegetables, and, sprinkled on meat, hide the taste of excess fat.

Mentha pulegium is pennyroyal, one of the few bitter-tasting mints, now used as a refreshingly scented ground cover. The specific *pulegium* < *pulex* Latin, flea, because the plant was used by the Romans as a fleabane. *Pulegium* entered Medieval

French as *poullieul*, and when the Normans finally brought the word to England, British ears and the vagaries of folk etymology could make only "pennyroyal" of the French.

∾

** OREGANO & SWEET MARJORAM**

Genus: Origanum < *origanum* Latin, Pliny's word for wild marjoram < *origanon* Greek, wild marjoram < *oros* Greek, mountain + *ganos* Greek, joy. The current English form was borrowed directly from Spanish, *oregano*. Italian is *origano*.

Family: Labiatae, the mint family

SPECIES
Origanum vulgare (Latin, common) is the oregano whose savour has become jaded for North American palates through its overuse on pizzas. But note that the dried, pathetic oregano usually sold in North American spiceries is actually a feeble Mexican variety of wild marjoram that has none of the vigour and dash of home-grown Italian *origano*.

Origanum majorana is sweet marjoram with a milder, though interesting flavour when the leaves are fresh and not dried. The Botanical Latin specific *majorana* < *amaracus* Latin < *amarakos* Greek, this plant < *marjami* Arabic, this plant < akin to a word in Dravidian, an Indian language.

∾

PARSLEY

Other common names: French curly parsley, Italian plain-leaved parsley

Genus: Petroselinum < *petroselinum* Latin, Pliny's word for rock-parsley < *petroselinon* Greek, parsley < *petros* Greek, a stone + *selinon* Greek, parsley, celery. The English version parsley < *persely* Middle English < *peresil* Old French < *petrosilium* Medieval Latin < *petroselinum* Latin.

I have to include here a delightful Greek nursery rhyme that has survived for almost three thousand years. It was said in a children's time-passing game like paddycake-paddycake. It does not look neat with the Greek letters transliterated and the

Petroselinum Macedonicum verum.
The true Parſley of Macedonia.

accents left off, but: *Pou moi ta rhoda; pou moi ta ia; pou moi ta kala selina; tade ta rhoda, tade ta ia, tade ta kala selina!* Translation: Where are my roses; where are my violets, where are my beautiful parsleys? Here are your roses; here are your violets; here are your beautiful parsleys. The children probably pointed, giggling, to various body parts as this innocent rhyme was chanted.

The Greek words for rock and stone, *petra* and *petros*, give rise to many current English terms: The Petrified Forest, where wood has turned to stone, petroleum (oil from a rock), saltpetre, the storm petrel (a bird), and even parrot. The given name Peter has the same origin, and is subject to the most egregious and niftiest pun in the New Testament, where Matthew reports Jesus saying to Peter (*Petros* in the Koine Greek of the New Testament): "Thou art Peter, and upon this rock (Greek: *epi tautei tei petra*) I will build my church."

Family: Umbelliferae, the carrot or parsley family

SPECIES

Petroselinum crispum (Latin, with close waves, said of the leaves) is often used raw as a garnish. *Petroselinum filicinum* (Botanical Latin, fern-like, of the leaf arrangement < *filix* Latin, fern) is Italian or Neapolitan parsley, whose young, tender leaves are sometimes cooked as a vegetable and are often used as a culinary herb.

ROSEMARY

Genus: Rosmarinus < *ros* Latin, dew + *marinus* Latin, of the sea < *mare* Latin, sea. Named Dew of the Sea because the plant grew wild on cliffs beside the sea in southern Europe.

Family: Labiatae, the mint family

SPECIES

Rosmarinus officinalis (Medieval Latin, of the herbalist's shop, *officina*) is the evergreen shrub whose fresh leaves are so richly aromatic that, where it grows wild on the Mediterranean coast, perfuming rocky hillsides, the scent of rosemary can be smelled ten kilometres out at sea, when the wind is southerly and rosemary is being harvested. Canadians usually grow this tender herb as a pot annual, starting new pots each year from stem cuttings, summering the pots in full sun, and wintering indoors, perhaps at a sunny kitchen window. The English word has no connection with rose or Mary, it is our old and comforting friend in the face of foreign words, folk etymology, which naturally insists on trying to find recognizable English roots in strange terms. Rosemary's spikey leaflets will bring to life meats other than lamb. It is the herb of memory, most famously in Shakespeare's *Hamlet*, where Ophelia says, "There's rosemary, that's for remembrance; pray, love, remember …"

∾

SAFFRON

See under *Crocus* in *Bulbs*.

∾

SAGE

See under *Salvia* in *Annuals*.

∾

SORREL

Other common name: French Sorrel

Genus: Rumex < *rumex* Latin, wild sorrel < *rumex* Latin, a slightly curved hunting spear, possibly because the leaves of sorrel looked with a small shield and the flower spike like a hunting spear (?)

Family: Polygonaceae, the knotweed family, named after its typical genus, Polygonum < *polys* Greek, many + *gonos* Greek, offspring, because of the many seeds; or + *gone* Greek, knee, because the knobbly joints of the stems looked like so many knee joints

SPECIES

Rumex scutatus (Latin, shield-like, of the leaves) is French Sorrel, whose shield-shaped leaves have a sour, lemony taste that adds zest to salads and cooked dishes when they are picked young and fresh. Sorrel sauce is a classic accompaniment of fish in French cuisine.

∾

TARRAGON

See under *Wormwood* in *Perennials*.

∾

THYME

Genus: Thymus < *thymon* Greek, thyme, the familar aromatic herb of cooking, but also used by the ancient Greeks to add a resinous perfume to burnt sacrifices, hence its root in the verb *thyein* 'to burn as a sacrifice to the gods.' This large genus has over 100 species, more than a dozen of which are used as herbs.

Family: Labiatae, the mint family

SPECIES

Thymus vulgaris (Latin, common) is the garden thyme of French cuisine with strong flavour and scent. *Thymus citriodora* (Botanical Latin, lemon-scented) is lemon thyme which adds an intense, citric savour to many dishes. Oberon says in *A Midsummer Night's Dream*, "I know a bank whereon the wild thyme grows," referring to *Thymus serpyllum* (Botanical Latin, creeping), which is pleasant and aromatic in the chinks of rock gardens or allowed to run along a summer terrace.

∾

WATERCRESS

Genus: Nasturtium < *nasus tortus* Latin, twisted nose, referring to the pungent taste of the plant's leaves. Note that the garden plants popularly called nasturtiums are classified in botany under the genus Tropaeolum. Note too that watercress has been reclassified as *Rorippa nasturtium-aquaticum*. Some say the genus name is from an Old East German common name for the plant. But it looks suspiciously like a contraction of a Vulgar Latin phrase like *ros ripae* Latin, dew of the riverbank, in reference to the habitual site of watercress.

Family: Cruciferae, the mustard family < *crux, crucis* Latin, cross + *ferre* Latin to carry, to bear. The cross they bear is the arrangement of their flower petals, usually four petals at right angles to one another, making the grouping resemble a little cross. A large family, the crucifers include many important vegetables; some like cauliflower and broccoli are considered anti-carcinogenic, others are nutritious like cabbage, brussels spouts, turnip, and radish.

SPECIES

Nasturtium officinale is now *Rorippa nasturtium-aquaticum*, but it is still peppery, tasty watercress to me, and still makes an excellent salad ingredient, if you can find it or grow it fresh. And watercress sandwiches are still *de rigueur* as munchables to accompany high tea—on a silver salver, please.

PERENNIALS

PERENNIALS

See *Black-eyed Susan Vine* in *Annuals*.

ACANTHUS

❦

ADONIS

Common names: Amur Adonis, Pheasant's Eye (British), Spring Adonis

Genus: Adonis < Greek, the name of a handsome youth in the myth told below, who spent half the year above ground living life on earth and then plunged into the underworld during the winter, much as this yellow-flowered perennial bursts through the frost-free loam early in spring, thrives for two months, then dies down to the ground until the following spring. *Adonis*, Greek < *adon* Sumerian, lord, ruler. A common Proto-Semitic title of respect, *adon* has a reflex in Biblical Hebrew. Compare Adonai as a name of God in the Old Testament and in the exalted syllables that open some of the great Hebrew blessings, *Barukh Attah Adonai Elohenu* "Blessed art Thou, O Lord our God."

Family: Ranunculaceae, the buttercup family, named after its typical species, the buttercup, Ranunculus < *rana* Latin, frog + *unculus* Latin, diminutive suffix, little, tiny. *Ranunculus* is Latin for froggie-woggie, because buttercups like damp ditches and moist dells just as frogs do.

The Latin word for frog is echoic; *rana* imitates certain calls of the male frog, just like ribbit-ribbit in English. Like us, the ancients were charmed by animal noises. One of the best and most playful amphibian imitations occurs in a comedy produced in 405 B.C. It is the famous Frog Chorus from Aristophanes's *The Frogs* and can be transliterated exactly from ancient Greek as: Brekekekéx-koax-koax!

SPECIES

Adonis annua (Latin, annual) blooms in the fall in English gardens where it is commonly called Pheasant's Eye. *Adonis aestivalis* (Latin, of the summer) blooms in the summer. The most familiar Adonises in North America are *Adonis vernalis* (Latin, of the springtime) which blooms a few weeks after the earliest one, *Adonis amurensis* (Botanical Latin, from the vicinity of the Amur River which forms part of the boundary between China and Russia).

THE MYTH OF ADONIS

Adonis was borrowed by the Greeks from Asian and Middle Eastern fertility myths where in his earliest guise as Sumerian Dumuzi or Akkadian Adon Tammuz, lord of agriculture, or Phrygian Attis or Egyptian Osiris, he represents a spirit of vegetation that is born and dies and is born again the next spring; and so Adonis personifies the cycle of the year, the changing seasons of spring growth, summer bounty, and winter decay.

The rite of Adonis involved women mourning by carrying around effigies of a dead youth. More striking, however, were the "gardens of Adonis" as the ancient Greeks called them. These were crockery pots of any quick-growing seed placed on roofs and beside fields and doors. The TV ads hawking clay figurines like the Chia Pet™ smeared with seed-containing paste are thus part of a tradition of fertility novelties reaching back more than six thousand years.

The Adonis myth was personalized by the Greeks. Instead of vegetation gods rising and sinking on some impersonal seasonal elevator, the Greeks imagined a handsome youth loved equally by two powerful goddesses who had to share him. Aphrodite, goddess of earthly love, had Adonis for the first part of the year, spring and summer. Persephone, love goddess of the underworld, claimed Adonis for the late fall and winter, and their rooty coupling took place deep in the chthonic realms below the cold earth.

It is clear that Adonis is a plant turned into a mythical person. In one of the stories about the death of Adonis, grizzled old Ares, Greek god of war, is jealous of Aphrodite's lust for the firm limbs of young Adonis. So Ares disguises himself as a wild boar and gores Adonis to death. As the youth exsanguinates, blood from his mortal wounds soaks a nearby flower, staining it red, thus accounting for red anemones. Other mythographers say that red anemones first sprang from Adonis's blood. Attis had blue blood, so when he expired, violets sprang up. Well, our perennial Adonis usually has yellow flowers. Perhaps Adonis had a cirrhotic liver from all that dandelion wine slurped on the meadowy flanks of sacred mountains?

∾

Other common names: Windflower, Pasque Flower, Poppy Anemone

Genus: Anemone < *anemone*, Greek, wind flower, literally 'daughter of the wind'< *anemos* Greek, wind. But *anemone* is perhaps an example of folk etymology in which the Greeks heard the name of a mythic Semitic hero, Naaman, and tried to shape it into a familar Greek word. Naaman was a vegetation deity like Adonis who also died annually, and from whose shed blood sprang forth the deep red *Anemone coronaria*. It is called Windflower because its plumed seeds parachute on autumn winds, borne aloft as they disperse.

Family: Ranunculaceae, the buttercup family; see *Adonis* entry.

ANEMONE

Pasque Flower

SPECIES

Canada Anemone
The Canada Anemone of our eastern provinces, *Anemone canadensis* (Botanical Latin, Canadian), can be cultivated in gardens, although it is an intrusive little rooter that quickly invades garden space set aside for showier plants. Better perhaps to enjoy its white flowers in the wild as nodding tokens of late spring. The Canada anemone likes wet meadows and is comfy on the banks of creeks and forest brooklets. But some Canadian nurseries do offer the seed, so please don't uproot mature plants from their wild sites.

Provincial Flower of Manitoba

The official flower of Manitoba is a prairie anemone, Pasque Flower, *Anemone patens* (Latin, spreading). Related to *patens* is the name of a little, flat, table bowl used by the ancient Romans to spread out raw vegetables or meatballs, and then borrowed by anatomists to name the roundish kneecap. The word is patella.

WORD LORE OF PASQUE FLOWER

Pasque is one Norman French spelling of the Old French word for Easter, *Pasches*, ultimately through ecclesiastical Latin *pascha* to Koine Greek *pascha* to Aramaic *paskha* to its origin in the Hebrew word for passover, *pesach*. But why Pasque Flower? Well, some ponderous tomes of botanical wisdom say it's because in England it blooms approximately around Easter (actually in England it's a different but related plant, *Anemone pulsatilla*). One very famous writer on plants, John Gerard, even claimed in *The Herball, or General Historie of Plants* published in 1597 that *he* named the plant due to its Easter flowering. But a bit of riffling through other, more objective ancient herbals uncovers the news that Pasque Flower gave a juice from its sepals that was used as a green dye for painted Easter eggs in many European countries. But Canadians brought up on the prairies know that downy seedheads of Pasque flowers being combed by a stiff breeze mean autumn is truly here and winter's but a stubbled field or two away.

Anemone blanda (Latin, pleasant) grows from little black tubercles stuck into October loam to bloom brightly next June. *Anemone pulsatilla* (Late Latin, literally 'little striker,' reference unknown) appears modest in bloom but produces gaily feathered seedheads that look great in dried bouquets. *Anemone coronaria* (Botanical Latin, used in a *corona* Latin, flower crown, garland) is native to Mediterranean shores and makes a usually hardy perennial in Canadian gardens.

❧

ASTER

Common names: Starwort (early aster name, 1578 A.D.), Michaelmas Daisy (1582). In England, some asters are called Michaelmas daisies. Michaelmas (St. Michael's Mass or Feast)

is an Anglican and Roman Catholic religious celebration of
Saint Michael which falls traditionally on September 29.

Genus: Aster < *aster* Latin & Greek, star, referring to the shape
of the flower

Family: Compositae, the daisy family < *compositus* Latin,
placed together. This largest family of flowering plants is named
after the compound flowers of its members. Small
florets of individual flowers make up large clusters or heads.

SPECIES

Aster novae-angliae (Botanical Latin, of New England). This
species of aster is native to eastern North America. Long ago,
British and European plant breeders collected samples and
seeds of *Aster novae-angliae* and used it to create garden varieties of asters, some called Michaelmas daisies. But, how apt is
this British common name? Michaelmas daisy. Well, it is an
aster, not precisely a daisy, although some daisies reside in the
genus *Aster*. Many varieties of this aster bloom at the end of
August, long before Michaelmas. You decide which name is
more apt. *Aster* is the Latin word for 'star' and refers to the
shape of the flowers in this genus, the star-like, radiating
arrangement of its petals. Manitoban hybridizers have produced
several asters suited perfectly to the Canadian Prairies.

North America has some 300 species of native asters that
hybridize freely, making precise identification a bit of a taxo-
nomic nightmare. But across our wild fields, often until first
snowfall, asters add a muted purple to the quilt of Canadian
autumn.

∾

Common name: False Spirea

ASTILBE

Genus: Astilbe < *a* Greek, not + *stilbos* Greek brilliant, glittering;
referring to the very small, inconspicuous white or pink flowers.
Some early wild species had pale, muddy-white flowers. Astilbe
is an old but unsatisfactory botanical name, because it is foolish

to name something for what it is not. This generic tells us little about the plant.

Family: Saxifragaceae, the rockfoil family, named after its typical genus, *Saxifraga* Latin, 'stone-breaker,' because it likes to root in patches of gravelly soil in rock clefts and is at home in limestone scree as well. Whether or not a cumulative effect of such rooting is to actually split a pebble is doubtful. It's more likely that early botanists simply noted its preferred site. Another suggested reason for the name stone-breaker is its use by early herbalists as a supposed remedy for gallstones and bladder stones. An infusion of certain plant parts drunk regularly was thought to dissolve the calculi. There is no medical evidence to support this claim. Today a doctor is more likely to suggest ESWL or extracorporeal shock-wave lithotripsy in which sound waves conducted through water disintegrate the stones and the debris is washed out of the system.

Back in nature's rock garden, an older English name for saxifrage is rockfoil, and not because it "foils" rocks. This foil is an Englishing of the French word for leaf, *feuille*, itself from *folium*, Latin, leaf.

SPECIES

Astilbe arendsii is the mother of most commercial hybrids whose dense plumes add fine pastel accents to Canadian borders. The father is often one of the exotic astilbes from China, Japan, and Nepal. The specific recalls Georg Arends of Ronsdorf (1862-1952) who worked on hybridizing astilbes for many years and with great success.

❧

BEE BALM OR MONARDA

Other common names of different members of the genus: Horse Mint, Lemon Mint, Wild Bergamot, Sweet Bergamot, Oswego Tea

Genus: Monarda < after Nicholas Monardes (1493-1588), a doctor and plant collector in Seville who took an early interest in plants from America brought back to Seville by Spanish sailors. In 1577 one of Monardes's books containing a description of some species in this genus was translated into English as *Joyfull Newes out of the newe founde Worlde*.

Family: Labiatae, the mint family < *labiatus* Botanical Latin, lipped, with a prominent lip < *labia* Latin, the lips. In botany, the labium is the lower lip of a flower with two lips. In flowers of the mint family, this labium is highly developed and enlarged.

SPECIES

Monarda didyma (Botanical Latin, twinned, occurring in pairs < *didymos* Greek, double, twofold, twin) is the North American Bee Balm, Sweet Bergamot, or Oswego Tea. The specific recalls its paired leaves and double stamens. Its northern range ends in Québec and Ontario. The scent of its flowers and leaves reminded an early describer of the Bergamot orange, a citrus tree native to the northern Italian city and province of Bergamo. The somewhat minty essence of Bergamot is prepared as a zest from the rind of this citrus and was used in the eighteenth century to scent snuff. Over the centuries, several members of the mint family have been called bergamot. Today bergamot leaves are still used as flavourings to add mintiness to fruit cups and preserves. The Oswego People of upstate New York made the famous mint tea by steeping young leaves in hot water.

   Monarda citriodora (Botanical Latin, smelling like citrus < *citrus* Latin, citron tree) is the wild, deliciously perfumed Lemon Mint of North America, which also makes a fragrant garden subject.

∽

Common names of related species: Blue Staggers, Dutchman's Breeches, Fumewort (Old English), Ladies' Lockets (Victorian English), Lyre Flower

BLEEDING HEART

Genus: Dicentra < Botanical Latin, with two spurs < *dis* Greek, twice + *kentron* Greek, spur. The generic name refers to the two spurs on each flower. The prime meaning of *kentron* is sharp point; compare *kentein* Greek, to prick. In Greek mathematics, a sharp-pointed compass was used to draw a circle, hence the easy borrowing by the Romans to give the Latin word *centrum*, the centre of a circle. Thus the Greek and Latin roots are bases for dozens of words in modern English like concentrate, eccentric, egocentric, epicentre, heliocentric, shopping centre.

Family: Fumariaceae, the fumitory family, named after its typical but now obsolete genus, *Fumaria*, based on a medieval common name, fumitory, from *fumus terrae* Latin, smoke of the earth, a reference to the feathery, fern-like leaves, which in large stands might by a stretched metaphor be said to resemble very vaguely smoke. By some botanists the plant is placed in the poppy family, Papaveraceae.

SPECIES

Dicentra cucullaria (Botanical Latin < *cucullus* Latin, little hood worn by a Roman child to fend off chilly weather, a reference to the shape of the flowers) is Dutchman's Britches or Breeches.

Another common name of gruesome mien is Blue Staggers, reminding us that the plant is toxic and contains alkaloids known to have killed cattle after a slow poisoning in which the poor beasts stumble, tremble, and gasp horribly for breath. *Dicentra eximia* (Latin, unusual, distinguished) is the bleeding heart with finely divided leaves and little plumes of the curious flowers. Unlike Dutchman's Britches, *eximia* blooms all summer.

Dicentra spectabilis (Latin, showy) is Bleeding Heart, a Japanese species, the subject of some hybridizing to produce spectacular perennials with pure white flowers.

None of the Dicentras work as cut flowers. The moment they are cut, they droop like moping crackheads deprived of a store to rob. Dicentra is a genus that tricks bumblebees. The stout little buzzers have short tongues that can reach the pollen, but the nectar is too far down the spurs, so bumblebees pollinate the plants which deny them any reward.

BUTTERFLY WEED

Other common names: Orange Milkweed, Showy Milkweed, Monarch Butterfly Plant, Pleurisy root

Genus: Asclepias < *Asklepios* Greek, god of medicine < *skalapazein* Greek, to entwine, as his symbol, the snake, winds around the caduceus. The Romans called him *Aesculapius*. In Greek and Roman art, he carries the caduceus, a winged stick encircled by a snake or two snakes. The caduceus was also the wand carried by a professional Roman messenger, the traditional sign of a herald who comes in peace to bring news.

But the stick encircled by the snake had been a symbol of magic healing in the ancient Near East for millennia. Long before Asklepios was ever depicted carrying the messenger's rod, the same caduceus—with one snake—was the magic wand of pre-Hellenic healers from whom the cult of Asklepios arose. Among the ancient Greeks, Asklepios was revered as the founder of medicine and achieved divine status as the god of healing. So popular was he in ancient Greece that shrines were devoted to his worship, where the sick or their relatives came, made an offering, and hoped for a specific cure to be revealed magically in dreams. Part of the ceremony in the chief temple of Asklepios at Epidaurus involved snakes kept by the attending priests.

The snake has been a worldwide symbol of healing, probably since the dawn of *homo sapiens sapiens*. The snake symbolizes fertility, rejuvenation (shedding its skin annually), keen sight, wisdom, and finally active health. The U.S. Army Medical Corps uses a two-snaked caduceus as its symbol. Their British equivalent, the Royal Army Medical Corps, uses a caduceus with one snake.

Family: Asclepiadaceae, the milkweed family, after its typical genus

SPECIES

Milkweed is *Asclepias tuberosa* with a specific from Botanical Latin that means 'growing from tubers,' thickened underground stems modified to store starch as food for the plant < *tuber* Latin, swelling < **tubh-* Indo-European root, be fat, be thick, swell, hence hundreds of kindred words in many IE languages, words like protuberant, thigh, thumb, tomb, tuberculosis, tumour, tumult, *tumulus* (mound), and turgid: all containing the idea of swelling. A tomb swells with the dead and with the earth piled above the burial site. Milkweed refers to the milky latex in the stems. Monarch Butterfly larvae feed on this white juice. It was called pleurisy root because aboriginal peoples of North America used it as part of a herbal treatment for certain inflammations of the lungs. Technically pleurisy is inflammation of the pleurae, membranes that line the thorax and partially enfold each lung. The pleural cavities may become filled with fluid in pleurisy. Asclepias has a deep taproot and thus resents being disturbed, so plant it where you will keep it. This wild beauty is the bearer of too startling an orange to be called a weed. It will bloom from seed in two years and stay for years to brighten a dry, sunny spot in the garden.

Asclepias speciosa (Latin, showy) is a western relative that will do well in gardens too.

CHINESE LANTERN

Other common names: Bladder Cherry (Yech! It sounds like something a urologist might wish to excise.), Cape Gooseberry, Ground Cherry, Winter Cherry

Genus: Physalis < Latin, bladder plant < *physa* Greek, bladder, referring to the large, papery calyx, of orange-red hue, which resembles a miniature Chinese lantern, named during the craze for chinoiserie, the inclusion of Chinese motifs in late nineteenth-century European design. The genus is native to the Caucasus and China as well as southern Europe, and was known early to European botany, grown in ancient Rome, mentioned in the writings of Dioscorides, a first-century A.D. Greek writer on the medical uses of plants. It is illustrated in a copy of his work published about 512 A.D.

Family: Solanaceae < *solanum* Latin, the black nightshade plant. The nightshade family also includes such important food plants as potatoes, peppers, tomatoes, and eggplants. In this family are *nicotiana tabacum*, common tobacco, as well as ornamentals like petunia and schizanthus, and poisonous plants like deadly nightshade.

SPECIES
Physalis alkekengi is the species with smaller lanterns. *Alkekengi* < *al kakanj* Arabic < *halikakabon* Greek, origin unknown. Chinese lantern spreads naughtily and overtakes

whole beds. So confine it early. A variety with larger lanterns was introduced from Japan in 1894 and dubbed *Physalis alkekegi franchetti*. The common name Bladder Cherry refers to the lanterns and the red berries. Cape Gooseberry is *Physalis peruviana* (Botanical Latin, of Peru).

∾

See in *Annuals*.

∾

Other common name: Honeysuckle

Genus: Aquilegia. Most dictionaries and most books about the origin of botanical names state that the generic Aquilegia stems from the Latin *aquila* 'eagle' from a supposed similarity of the flower spur to the claw of an eagle. Most dictionaries are wrong. Now the adjective from *aquila* is *aquilinus*, eagle-like. We know it from long, hooked noses, which English may term aquiline. For Aquilegia to have anything to do with eagles it would mean that for this one and only time in all the history of borrowing Latin words into English and French an *n* becomes a *g*. Well, it never happened. It's against all the many rules of consonantal transmutation in the Romance languages.

   This spurious origin of Aquilegia is a choice example of a learned dictionaryism, concocted over a dusty volume one midnight by some hapless word-nut and lazily passed down through the years from dictionary to dictionary without a single text that illustrates or proves or suggests the likelihood of the etymology. There is an etymologically sensible origin, it so happens, in the ancient Roman adjective *aquilegus*. As an adjective it means 'drawing water,' just as the

CHRYSANTHEMUM

COLUMBINE

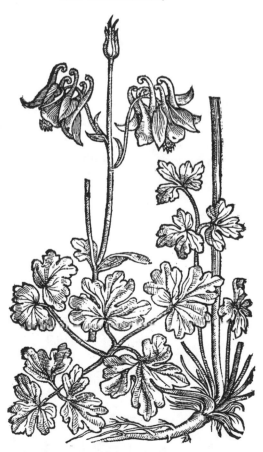

Aquileia rubra.
Red Columbines.

spur in the flower of the columbine collects its sweet droplet of nectar. The source is apt, possible, clear, and compelling to reason. But you won't find that derivation in any dictionary I know of.

Family: Ranunculaceae, the buttercup family, named after its typical species, the buttercup, Ranunculus < *rana* Latin, frog + *-unculus* Latin, diminutive suffix, little, tiny. *Ranunculus* is Latin for little frog, because buttercups like damp ditches and moist dells just as frogs do.

SPECIES

Our common eastern columbine is *Aquilegia canadensis*, whose ruby bells chime with nectar sweet enough to summon hummingbirds, moths, bumblebees, and butterflies: all long of tongue and deft of theft, when it is a question of the honeyed drop at the end of each flower spur. Honeybees who have no long tongues to probe simply chew through the spur to lap up the nectar.

Out west is British Columbia's blue columbine, and high up in the Rockies blows the yellow columbine. Almost 70 species of Aquilegia thrive in North America. The state flower of Colorado is *Aquilegia caerulea* (Botanical Latin, sky-blue, dark-blue), Rocky Mountain Columbine. Many, floriferous hybrids have been produced like the McKana Giants and McKana Improveds. Wild or hybrid, this honeysuckle loses its vitality after three or four years and should be replaced then.

WORD LORE

The silliest and most spurious etymology hovers about the word *columbine*. Now check out this real doozie. *Columba* is the Latin word for dove. And, yes, *columbinus* is the adjective meaning 'of or like a dove or pigeon.' Fine. Why is the flower so named? Well, let's quote one edition of *Webster's New International Dictionary*: "from the fancied resemblance of its inverted flower to a group of five pigeons." Isn't that cute? Their teeny pigeon heads turned inward to drink at the nectar droplet in the spur of the flower. What a load of codswallop! Just picture it. A person is walking through a meadow, notices the flower, examines it, and says, "Just like five doves turned inward to drink!" Odd. When I examine a columbine, as I so often do each morning while skipping down my garden path, I see five stegosauruses doing the macarena on a proton.

Once again, a commonsensical and cogent hypothesis exists. The Latin adjective *columbinus* gave rise to many female proper names in Romance languages, the most familiar being French *Colombine* and Italian *Colombina.* In the *commedia dell'arte*, the rollicking, improvising farce of the late Italian Renaissance, Colombina was a stock character, the mistress of Harlequin, in early days played by a hussy in outrageous jester's motley, played for laughs. In fact, one particular joke, used even later by the French writer of comedies, Molière, involves setting up her entrance. She is praised for her modesty, her chastity, the demure manner of her dress. Then on comes Colombina, a flirt and a tart dressed as gaudily as the company's meager resources will allow. Could this flower not have been named for her? In all European languages, it is quite common to label flowers with female names and properties. Think of marguerite, of maidenhair fern, of *Belle Angélique* et cetera. Far-fetched? It may be useful to recall that, in the Victorian Language of Flowers, the columbine was the symbol of folly, because its flower looked all loose and jangly like a jester's cap and bells. In the very exuberant vocabulary of Renaissance Italian, *una colombina* came to mean a woman who was one's ladylove. So the word referred to a lovely woman and was easily transferred to a lovely flower. No need at all to conjure five frumpish pigeons cooing petulance and sipping nectar.

Other common names: Gloriosa Daisy, Black-eyed Susan

CONEFLOWER

Genus: Rudbeckia < Olof Rudbeck, Swedish polymath (1630-1702), whose son was a friend of Linnaeus, the great revisor of botanical nomenclature. Linnaeus named this genus to honour both the father and the son. These coneflowers, all North American daisies, have yellow petals with a dark centre formed like a raised cone. Annual, biennial, and perennial forms are found in this genus.

Family: Compositae, the daisy family < *compositus* Latin, placed together. This largest family of flowering plants is named after the compound flowers of its members. Small florets of individual flowers make up large clusters or heads.

SPECIES
Rudbeckia hirta (Latin, hairy, said of the stems and leaves) is the bright yellow wildflower of woods and meadows, flourishing from the Canadian Prairies to the Maritimes. It does not bloom as profusely as *Rudbeckia gloriosa* hybrids, the bountifully floriferous gloriosa daisies, said to be perennial but usually winter-killed in Canadian gardens, only however after blooming their hearts out the first summer. *Rudbeckia fulgida* (Latin, shining, but literally 'yellow as lightning') is called orange coneflower. An outstanding cultivar called "Goldsturm" makes a perfect bedding plant, giving bright, fat mounds of Black-eyed Susans.

∾

CORAL BELLS

Other common name of a related species: Alum root

Genus: Heuchera < after Johann Heinrich von Heucher (1677-1747), professor of medicine and enthusiastic amateur botanist at the University of Wittenberg, Germany.

Family: Saxifragaceae, the rockfoil family, named after its typical genus, *Saxifraga* Latin, 'stone-breaker,' because it likes to root in patches of gravelly soil in rock clefts and is at home in limestone scree as well.

SPECIES
Heuchera sanguinea (Latin, blood-red, after the vibrant colour of this Mexican species' flowers) is the common Coral Bells, with many varieties and cultivars producing pink, white, and other flower colours. *Heuchera americana* of the southwestern United States is the pioneer herb, Alum root, named because an astringent like alum was obtained by processing the roots and used as a styptic lotion to check the bleeding of small wounds and razor nicks from careless shaving.

∾

Older common name: Lily Asphodel

DAY LILY

Genus: Hemerocallis, Botanical Latin < *hemera* Greek, day + *kallos* Greek beauty, in reference to the beautiful flowers that each last but a day. Compare the same Greek root for *day* as it appears slightly disguised in the adjective *ephemeral* (*epi* Greek, on + *hemera* a day), said of anything transitory or of short duration. In its adjectival form *kalos*, the Greek root for beauty has many derivatives in modern English: calisthenics (literally, beautiful strength), calligraphy (beautiful writing), calliope (beautiful voice; Calliope was the Greek muse of eloquence and epic poetry), kaleidoscope (viewing device to observe beautiful forms), and even the rare adjective, callipygian (having beautifully formed buttocks).

Family: Liliaceae, the lily family < *lilium* Latin, lily. The Latin word is related to *leirion* Greek, lily, and is one of the ancient Mediterranean flower words. Both Latin and Greek appear to have borrowed the word early from a Coptic form *hleli*, itself a variant of *hreri*, all stemming from ancient Egyptian *hrr*, lily. The lily family is so huge that subfamilies have been formed, and the day lily has its own little group here called Hemerocallidaceae.

SPECIES

More than eleven thousand cultivars of this splendid, disease-free, easy-to-grow perennial are available. But the old species make wonderful subjects too. Native to Asia are the fragrant lemon-yellow trumpets of *Hemerocallis lilioasphodelus*, where the specific recalls the old name of Lily Asphodel < *asphodelos* Greek, a lily. Here is an example of how plant names get tossed around in history as they are borrowed from language to language. The asphodel in Middle Dutch was *de affodil*, borrowed into English to produce daffodil!

❧

Lilium non bulbosum Phœniceum.
The Day-Lillie.

DELPHINIUM

Other common name: Larkspur, from the little spurs on the flowers and the fact that the tall spikes of florets attract birds

Genus: Delphinium, Latin < *delphinion* Greek, little dolphin, the ancient Greek name for a different species, *Consolida ambigua*, whose flower the Greeks thought resembled the head of a young dolphin < *delphis* Greek, dolphin.

Family: Ranunculaceae, the buttercup family, named after its typical species, the buttercup, Ranunculus < *rana* Latin, frog + *-unculus* Latin, diminutive suffix, little, tiny. *Ranunculus* is Latin for little frog, because buttercups like damp ditches and moist dells just as frogs do.

SPECIES

Delphinium elatum (Latin, tall) is the giant candle delphinium whose excellent strains include Pacific, Blackmore & Langdon, and Wrexham Hybrids. One delightful old species is *Delphinium zalil* from Afghanistan where the specific names a yellow dye extracted from the lemony flowers and used to colour silk.

ERIGERON

Common names: Fleabane (from the medieval practice of strewing the floors of dwellings with the whole, crushed plant, supposed to keep fleas from entering a building and to kill the ones that fell off those who entered, man or beast)

Genus: Erigeron < *erion* Greek, wool + *geron* Greek, old man; thus woolly-haired old man, in reference to the white-haired seedhead of many species

Family: Compositae, the daisy family < *compositus* Latin, placed together. This largest family of flowering plants is named after the compound flowers of its members. Small florets of individual flowers make up large clusters or heads.

SPECIES

This delightful daisy provides two or three years of purple and pink flowers, some with yellow centres. But the species are attractive too; for example, *Erigeron aurantiacus* (Botanical Latin, orange-coloured) from Turkestan is a perky little orange-red daisy for the border.

∾

See under the family subsection in the *Castor Bean* entry in *Annuals*.

∾

Old common names: Bitterwort, Felwort (*feld* Old English, field + *wyrt* Old English, plant), Baldmoney (origin unknown)

Genus: Gentiana < named after King Gentius of Illyria who discovered the medical uses of the roots of yellow Gentian around 500 B.C. Illyria was an ancient kingdom on the east coast of the Adriatic Sea comprising what is now northern Albania and Montenegro. See *Word Lore* below.

Family: Gentianaceae, the gentian family, perennial herbs of worldwide distribution including species from New Zealand, the Himalayas, China, Tibet, Europe, the Caucasus, and North America.

EUPHORBIA

GENTIAN

SPECIES

Gentiana lutea (Latin, yellow, of the flowers) is native to Europe where its roots still provide gentian bitters or brandy, a supposed digestive and appetite stimulant. One of the best is the Swiss schnapps called Enzian, which is also the German word for gentian.

But the traditional colour of the flowers in this genus is a characteristic deep blue, gentian-blue. *Gentiana acaulis* (Botanical Latin, not-stemmed) is Trumpet Gentian with blue flowers. *Gentiana pneumonanthe* (Botanical Latin < *pneumon* Greek, lung + *anthos* Greek, flower) is Marsh Gentian or Lung Flower, from its supposed efficacy in easing breathing complaints. The most spectacular flower blooms from the hard-to-grow biennial *Gentiana crinita* (Botanical Latin, with hairs)

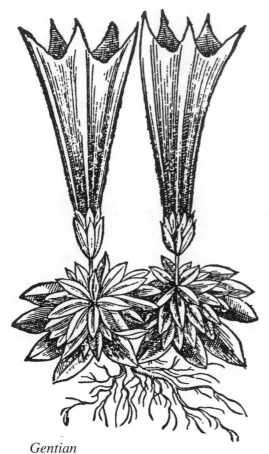

Gentian

or Fringed Gentian whose purply-blue petals curve voluptuously and end in beautiful fringed edges. Another fine blue gentian is *Gentiana septemfida* (Latin, sevenfold, said of the seven-part corolla).

WORD LORE

Most dictionaries, including the current *Oxford English Dictionary*, say that gentian owes its name to Gentius, relying for this assertion on two ancient authors. It will be of interest, I hope, to see the actual passage that gives us our only clue about the origin of the flower word *gentian*. It appeared in *The Natural History*, a primitive kind of encyclopedia of science written by Pliny the Elder, a working lawyer and lifelong public servant of Roman emperors. Born in 23 A.D. Pliny crammed fact-mongering into a life of imperial offices. He died from poisonous fumes while investigating the eruption of Mount Vesuvius in 79 B.C., the same volcanic disaster that buried the little town of Pompeii. While Pliny was not a scientist and was a trifle gullible, he did have a curiosity about the natural world that preserved many ancient bits of lore and knowledge. From Book 25, Section 34, of his *Natural History*, here is Pliny's entry on which, for the next two thousand years, dictionary-writers would depend for an explanation of the origin of the name of the gentian flower:

> *Gentianam invenit Gentius rex Illyriorum, ubique nascentem, in Illyrico tamen praestandissimam … acquosis montibus subalpinis plurima. Usus in radice et suco. Radicis natura est excalfactoria, sed praegnantibus non bibenda.*

> Gentius, king of the Illyrians, discovered the gentian, a plant that grows nearly everywhere but does best in Illyria … Gentian grows in great abundance on moist elevations in the Alpine foothills. The root of the plant and its juice are the parts most used. The essential virtue of the gentian root is that it warms the body.

However infusions of the root or juice should never be drunk by pregnant women. [Author's translation]

Only one other contemporary author gives the Gentius story, and he was Pedanios Dioscorides who flourished in the middle of the first century A.D. as an army doctor, a Greek in the Roman army, with some claim to having written the first, valuable pharmacopoeia. Entitled *De materia medica* it was the most copied and revered guide to the medicinal use of plants until the end of the Middle Ages. Dioscorides may have borrowed the gentian story from Pliny, or vice versa.

Uses

The bitter root of gentian species has been much employed in herbalists' decoctions. From the year 1597 A.D., here is John Gerard in *The Herball or General History of Plants*, writing of "the vertues of Felwoort," that is, gentian: "It is excellent good, as Galen saith, when there is need of attenuating, purging, clensing, and removing of obstructions, which qualitie it taketh of his extreme bitternesse … This is of such force and vertue, saith Pliny, that it helpeth cattell which are not onely troubled with the cough, but are also brokenwinded. The root of Gentian given in pouder the quantitie of a dramme, with a little pepper and herbe Grace mixed therewith, is profitable for them that are bitten or stung with any manner of venomous beast or mad dog."

❧

GERANIUM

Other common names: Crane's Bill, Herb Robert

Genus: Geranium, sixteenth-century Botanical Latin < *geranion* Classical Greek, the name of the flower < *geranos* Greek, crane. The seed pods of geraniums have long, slender beaks that do look like the bills of cranes. The Greek word has the same ultimate Indo-European origin as *crane*, the English bird name. IE **gru* 'crane' is echoic, from the characteristic call of a crane. It gives later words meaning crane like *grus* Latin, *Kranich* German, *garan* Welsh, and *cran* Old English.

Family: Geraniaceae, the geranium family. Note that also in this family are the so-called florists' geraniums that originated in South Africa, widely used for summer pot and bedding plants in North America. Known by some as Stork's Bill, they have been given their own genus, Pelargonium, derived from *pelargos* Greek, stork, a reference to their fruits that resemble the bill of a stork.

SPECIES

Herb Robert

Of the dozens and dozens of perennial geraniums available, we shall instead single out one little wildling, an annual raised from seed sown where the plant is wanted, Herb Robert, *Geranium robertianum*. Also known as fox geranium, garden-gate, red robin, or red shanks, this merry denizen of woods and damp places can be seeded in a similar garden site, but it does not like soil that is too rich. One poopy reference text describes this woodland wonder as "a sticky low geranium with small reddish-purple flowers." Harumph! I collected seeds years ago whilst traipsing through some open boscage and the next year found it a valued addition to a small section of Canadian wild flowers I once had in a garden left behind long ago in a move. Brought from Europe perhaps as seed hidden in a cloak, this aggressive geranium has colonized many places in Canada from our West Coast all the way to Newfoundland. It is certainly not "low," often reaching half a metre in height. Given a warm spring, herb Robert blooms in Canada from April into early summer. The Robert of the herb is probably Saint Robertus (1027-1111 A.D.), founding abbot of the Benedictine monastery at Molesme in Burgundy.

❧

GEUM

Other common names of various species: Avens, Chilean Avens, Creeping Europe, Herb-bennet, Water Avens

Genus: Geum < Latin, Pliny's word for this plant, adopted into botany by Linnaeus < *gaion* Doric Greek, this plant < *gaios* Doric Greek, of the earth < *gaia* Greek, land, country, earth

Family: Rosaceae, the rose family

SPECIES

Modern garden geums with their gaudy double flowers are
chiefly offspring of *Geum chiloense* (Botanical Latin, of Chile).
But some of the old species are attractive subjects for moist borders in the shade where they add needed colour. One such
is *Geum rivale* (Latin, growing by streams < *rivus* Latin, brook,
stream) or Water Avens (*avence* Old French < *avencia* Medieval
Latin, origin unknown), a modest delight with its pink petals
and purple calyx. Incidentally, our English word *rival* was
originally a friendly word for neighbours who were *rivales*,
that is, living on the same stream, perhaps across the riverbank
from each another, at first in peace, and then with spats and
neighbourly tiffs, developing a more modern sense of rivalry.

Herb-Bennet

Geum urbanum (Latin, growing in town and city < *urbs, urbis*
Latin, city) named by Linnaeus is herb-bennet, with yellow
flowers, a staple of medieval herb gardens where herb-bennet
roots were a spicy stand-in for cloves. A hot tea of herb-bennet
leaves was a popular, medieval anti-constrictive, being
pronounced effective against "stitches and griefe in the side."
Wood Avens or herb-bennet entered English < *herbe beneite*
Old French < *herba benedicta* Medieval Latin, blessed herb,
so-called because a sprig of the leaves worn on the coat was
said to keep the devil away from the wearer.

Other common name: Marsh Mallow

HOLLYHOCK

Genus: Althaea (frequently misspelled Althea) < *althaia* Greek,
marsh mallow < *althos* Greek, healing medicine, referring to the
curative properties of the mucilage in the roots

Family: Malvaceae, the mallow family < *malva* Latin, the
mallow plant, with the Latin word akin to *malakos* Greek, soft,
downy, and to *malache* Greek, mallow plant. The family
contains many garden subjects in its 50 genera and some 1,200
species.

SPECIES

Marshmallow was originally made from a mucilage extracted from the root of a hollyhock, the marsh mallow plant, *Althaea officinalis*, the specific reminding us that the plant extract was sold in the medieval herbalist's shop (*officina*) where this mucilage was considered a potent heal-all. The modern confection used in marshmallow sauce or partially melted over a camp fire on summer outings is strictly sugar, starch, and gelatine.

Elizabethans liked to cluster hollyhocks around the gates of cottages, while contemporary perennial beds in Canada contain cultivars of *Althaea rosea* (Latin, pink), native to China, where it was grown for more than a thousand years before being introduced into English gardens in 1573 and becoming immediately popular.

WORD LORE

Hollyhock (holy + *hoc*, a word for mallow in Old English) is an Englishing of its Late Latin name, *malva benedicta* 'blessed mallow,' blessed because of its medicinal properties. By 1625 poet and playwright Ben Jonson was writing of "King's-Speare holy-hocks."

USES

The mild mucilage of this hallowed mallow was often taken as an emollient laxative, being mentioned for this purpose by the Roman poets Horace and Martial. In the Middle East, the roots were boiled and then fried with onions and oil as a subsistence food in times of cereal-crop failure. Ancient Arabic medicine recommended a poultice of mallow mucilage to help inflammations. For centuries French druggists made lozenges of marshmallow paste, *pâté de Guimauve*, to treat coughs and hoarseness. Old French herbalists also boiled the flowers in oil and water as one of the ingredients in *tisane de quatre fleurs*, a gypsy tea of four flowers, used to soothe a raw throat.

Other common names: Funkia, Plantain Lily

HOSTA

Genus: Hosta < named after Nicholaus Tomas Host (1761-1834), a doctor in the imperial service of Austria who was an amateur botanist and who introduced some species into the royal gardens at Vienna. Host wrote of this plant a little earlier than a Prussian botanist named H.C. Funck (1771-1839), hence the change of the generic from Funkia to Hosta.

Family: Liliaceae, the lily family < *lilium* Latin, lily. The Latin word is related to *leirion* Greek, lily, and is one of the ancient Mediterranean flower words. Both the Latin and Greek words hark back to a Coptic form *hleli*, itself a variant of *hreri*, all stemming from ancient Egyptian *hrr*, lily. The lily family is so huge that subfamilies have been formed, and hosta has its own, Hostaceae, formerly Funkiaceae.

SPECIES

Many variegated hosta cultivars now throng the marketplace, with wavy-margined white and green leaves, ungainly blobs that look like uppity cabbages. Plant centres are brazen enough to recommend these splorps of greenery for bedding rows. I do know one feisty greenthumb who grows her own "ugly patch," a bed to which she humorously consigns the less winsome mutants of modern botanical gene-tinkering. Here is the deserved site for these horrid, squat hostas. Try instead some of the truly bold Chinese and Japanese species like *Hosta sieboldiana* whose silvery-green and bluish leaves form a noble mound. Phillip Franz van Siebold (1796-1866) was a German physician who collected plants in Japan and introduced many to European gardens. Easily divided, hostas prefer moist, hot shade and burgeon rampantly when richly manured. The tallest is *Hosta undulata* (Latin, wavy, said of the leaf margins). Strongly ribbed, heart-shaped leaves of pleasing formality are the highlight of *Hosta plantaginea* (Botanical Latin, resembling plantain leaves). The Royal Botanical Gardens at Hamilton, Ontario, have a wonderful collection of giant Japanese hostas planted down the slope of a densely shaded hill.

IRIS

Iris Anglica maior bulbosa latifolia.

Other common name: Flags

Genus: Iris, Latin < *iris, iridis* Greek, rainbow, then personified as a goddess and charged by Zeus with carrying messages from the immortals to humans across a polychrome sky bridge, the rainbow. Every time they saw a rainbow, the ancients thought a god was sending a private word to some mortal.

Family: Iridaceae, the iris family, a vast clan of 55 genera and more than 1,400 species, many gardenworthy

SPECIES

Of the hundreds of species and thousands of hybrids, we have space only to mention our favourite, Japanese iris, *Iris japonica*, and to say: find it, see it, plant it, enjoy a glimpse of paradise, for this beauty is indeed a message from the gods.

WORD LORE

Medicine did not borrow the word until 1721 when the Danish anatomist Jacob Benignus Winslow (1669-1760) first applied the word *iris* to the thin circular membrane with its central aperture that forms the pupil of the eye. Winslow used *iris* because of the varied colours of the membrane.

❧

JACOB'S LADDER

Genus: Polemonium < *polemonion* Greek, a plant associated with Polemon of Cappadocia

Family: Polemoniaceae, the Jacob's Ladder family, named after its typical genus, polemonium, a perennial herb with deep-blue flowers. The plant was named after one of medicinal use mentioned in the writings of the scholar Polemon of Cappadocia, who flourished around 180 B.C. as a writer of travel guides and antiquarian studies in which he delighted readers by constantly correcting his predecessors. A familiar summer annual, the Cup-and-Saucer vine, Cobaea, belongs in this family.

SPECIES

Polemonium caeruleum (Latin, sky-blue, deep blue, of the flowers) is Jacob's Ladder. The common name arose because the long leaf with tightly spaced leaflets vaguely resembles a ladder, and the nearest referential ladder to hand was the story of Jacob's dream about the gate of heaven in the Old Testament, Genesis 28:12: "And he dreamed, and behold a ladder set up on the earth, and the top of it reached to heaven: and behold the angels of God ascending and descending on it." Modern Biblical scholarship considers this vision the garbled folk memory of a nearby ziggurat, a stepped-mound building with ladder-like access-steps running up along its outer walls. Jacob is then quoted: "How dreadful is this place! this is none other but the house of God, and this is the gate of heaven." Genesis reports that Jacob called the place Beth-el (Hebrew, house of God). Scholars believe this clinches the ziggurat reference, since a *beth-el* was a temple, and a ziggurat-like temple may have stood at the site early in the Canaanite period.

Other common names: Betony, Bishopswort, Woundwort (*wyrt* Old English, plant)

Genus: Stachys < *stachys* Greek, spike, referring to the flowerhead.

Family: Labiatae, the mint family < *labiatus* Botanical Latin, lipped, with a prominent lip < *labia* Latin, the lips. In botany, the labium is the lower lip of a flower with two lips. In flowers of the mint family, this labium is highly developed and enlarged.

LAMB'S EARS

SPECIES

Stachys lanata (Latin, woolly) is Lamb's Ears covered with soft, white, woolly hairs and part of every comprehensive English cottage garden.

Stachys officinalis is Wood Betony, from the medieval physic garden, hence the specific which indicates that it was sold in a herbalist's shop or *officina*. Betony entered English < *bétoine* Old French < *betonica* Street Latin, and the name Linnaeus gave one species < *Vettonica* Classical Latin, Pliny the Elder's term in his Natural History, possibly from the name of an Iberian people, the Vettones. Some species are native to Spain. Betony was highly esteemed in times past. The chief physician of Augustus Caesar, emperor of Rome, was a herbalist named Musa who wrote a long treatise about the medicinal wonders of betony, in which he listed forty-seven diseases that might be cured by an infusion of the dried plant parts. As one writer put it, "Betony was once the sovereign remedy for all maladies of the head." The *Medicina Britannica* (1666) raves about "the most obstinate headaches cured by daily breakfasting for a month or six weeks on a decoction of Betony made with new milk and strained." Another once-renowned headache cure was Rowley's British Herb Snuff, which contains among other necessaries dried betony leaves.

Marsh Woundwort is *Stachys palustris* (Latin, of the swamp) once widely used as a vulnerary, a plant good for stopping infection in a wound. Many medieval vulneraries often turn out to have some antiseptic and antibiotic properties. The Latin word for wound is *vulnus*, hence the prime meaning of vulnerable, able to be wounded.

Stachys sieboldi is a vegetable, sometimes called Chinese artichoke, knotroot, or the Japanese name, *Chorogi*. Phillip Franz van Siebold (1796-1866) was a German physician who collected plants in Japan and introduced many to European gardens. Chinese artichoke, where available, is often sold under its French name, *crosne*, dubbed from the little village of Crosne southwest of Paris, where this Stachys was first grown as a commercial crop. It is one of the oddest vegetables in shape, the underground tubers looking like dirty, damaged pearls on a string.

See entry in *Herbs*.

LAVENDER

❧

See entry in *Bulbs*.

LILY

❧

Other common names: Bethlehem Sage, Blue Lungwort, Jerusalem Cowslip

LUNGWORT

Genus: Pulmonaria < *pulmo, pulmonis* Latin, lung. The now discredited Doctrine of Signatures held that a beneficent deity had marked plants useful for man with little clues or signatures. The darkly mottled leaves of some lungwort species resembled diseased lung tissue; ergo, a plant with such a signature must be an effective remedy against pulmonary disorders. Splotches on the leaves were also said to resemble flecks of sputum and phlegm hawked-up from diseased lungs.

Pulmonaria or Lungwort

Family: Boraginaceae, the borage family, a small group of European and Asian herbs grown for red dye from their roots and for their blue flowers, which are still used to garnish salads. Borage leaves are used to flavour punches and cocktails. Borage < *bourrache* Old French < *borrago* Medieval Latin < *abu 'araq* Arabic, father of sweat, because early Arabic medicine used borage as a sudorific or diaphoretic, an agent that causes sweating.

SPECIES

Pulmonaria angustifolia (Botanical Latin, narrow-leaved) gives spring colour to a shady garden nook, with flowers that open pink and turn sky-blue. Bethlehem Sage is an old English name for *Pulmonaria saccharata* (Botanical Latin, sugared), the specific describing the leaves that seem to be sprinkled

with sugar. Both these species and others were sold in medieval herb shops as a physic for the lungs under the name of *Pulmonaria officinalis*.

Uses

The young leaves of some lungworts are part of the flavouring in some vermouths.

❧

MEADOW RUE

Genus: Thalictrum < *thaliktron* Greek name for a plant like Meadow Rue, but note how the Greeks thought this delightful genus a cheerful, happy plant, not at all rueful. The generic is related to *thalia* Greek, abundance, good cheer, in turn akin to *thallos* Greek, young sprout, green shoot and *thallein* Greek, to grow luxuriantly, to thrive.

Family: Ranunculaceae, the buttercup family, named after its typical species, the buttercup, Ranunculus < *rana* Latin, frog + *-unculus* Latin, diminutive suffix, little, tiny. *Ranunculus* is Latin for little frog, because buttercups like damp ditches and moist dells just as frogs do.

SPECIES

Rue came into English first as an Old French form of the plant's Latin name, *ruta*, itself from Greek *rhute* (original sense unknown). But this second *rue* was influenced by a totally unrelated word for sadness, *rue*, which already existed in the Old English wordstock. Because the flowers hang down, this lovely plant was early saddled with the mournful label of rue, from *hreow* Old English, sorrow, regret, sadness. As any gardener who has grown it will report, the lacy, ferny foliage of meadow rue sets off the cloud of purple and yellow flowers in a most uplifting manner. Calling this plant *rue* is like calling a tulip *widow's mope*. Quite unacceptable. *Thalictrum aquilegifolium* (Botanical Latin, with leaves like columbine) is the species most familiar to Canadian greenthumbs, but by growing a few other species, one can have rue in blossom from late spring to early fall, after waiting two years if one starts it from seed.

❧

Other names: Wolf's bane, Autumn aconite, Badger's bane

The root of monkshood has been mistaken for horseradish. It is not, but contains an alkaloid so toxic it was used in the middle ages to poison badgers and wolves, hence the other common names, wolf's bane and badger's bane. Use extreme caution when handling the roots and other parts of this plant. Don't let juice get into scratches and the little abrasions gardeners are heir to, or there won't be any heirs.

Genus: Aconitum, Latin < *akoniton* Greek, the monkshood plant. The word looks to be compounded of *a* Greek, not + *konis* Greek, dust. But in what manner is this plant dustless? The answer is lost. I suggest that *akoniton* is a dialect variant or mangling of the other word that ancient Greeks used for this plant, *lykoktonon* 'wolf-killer.'

Family: Ranunculaceae, the buttercup family, named after its typical species, the buttercup, Ranunculus < *rana* Latin, frog + *-unculus* Latin, diminutive suffix, little, tiny. *Ranunculus* is Latin for little frog, because buttercups like damp ditches and moist dells just as frogs do.

SPECIES

The upper petals of a monkshood flower are fused to make an enclosure that looks like the cowl worn by medieval monks. This perennial can reach well over a metre in height in most gardens. *Aconitum napellus* (Latin, little turnip, from the shape of the root) is the common English monkshood. The plant used to poison wolves in Europe for more than two thousand years is *Aconitum lycoctonum* (*lykos* Greek, wolf + *ktonos* Greek, killing).

∾

See under *Poppy* in *Annuals*.

∾

MONKSHOOD

CAUTION: POISON

ORIENTAL POPPY

PEONY

Other common names: Chinese peony, fern-leaved peony

Genus: Paeonia, Latin < *paionia* Greek, the flower sacred to *Paion* Greek, Healer, court physician to the Olympian gods and godlets. Then the word was attached to Apollo as a by-name, since he was the god of healing. *Paiein* is a Greek verb meaning to strike, to touch so as to heal. A song of praise to Apollo was a paean.

Family: Paeoniaceae, the peony family

Peony or Paeonia

SPECIES

Paeonia officinalis is the peony of the ancients, sold in medieval herbalists' shops or *officinae*. This plant, native to a broad area from France to Albania, was ringed with superstition and magic long before Athens was even a gleam in the eye of Pericles. This peony with beautiful, single, dark crimson or white flowers was said to have fallen from the moon. Herbalists said that peonies glowed deep in the night, protecting shepherds and their flocks from injuries and devilish spirits. One early English writer claimed it was an efficient cure for madness: "If a man laieth this wort [plant] over the lunatic as he lies, soon he upheaveth himself whole."

Most gardeners grow Chinese peony, *Paeonia lactiflora* (Botanical Latin, with milk-white flowers) or one of its many, variously coloured hybrids.

Common names for prostrate forms: Creeping Phlox, Ground Pink, Moss Phlox, Moss Pink

PHLOX

Genus: Phlox < Greek, a flame, ancient name for a red-flowered plant not necessarily of this genus

Family: Polemoniaceae, the Jacob's Ladder family, named after its typical genus, polemonium, a perennial herb with deep-blue flowers. The plant was named after one of medicinal use mentioned in the writings of the scholar Polemon of Cappadocia, who flourished around 180 B.C. A familiar summer annual, the Cup-and-Saucer vine, Cobaea, belongs in this family and so does the garden perennial Jacob's Ladder.

SPECIES

Canada Phlox or Wild Blue Phlox is *Phlox divaricata* (Botanical Latin, spreading). It is native to eastern Canada but has a lovely cultivar, *Phlox divaricata laphamii*, with soft blue or white flowers in late spring. There are dozens of startling upright garden phloxes with spectacular panicles, but my favourites are the creeping forms that mat across a bed to give a late spring mound composed of thousands of vivid purple and blue flowers. The most familiar in garden centres are hybrids and crosses based on *Phlox stolonifera* **x** *subulata*. One bears stolons that grow across the ground rooting at each node, while *subulata* (Botanical Latin, shaped like an awl, said of the leaves < *subula* Latin, awl, any pointed tool used to poke holes in leather or wood).

WORD LORE

Phlox, phlogos Greek, flame, has relatives in English, some admittedly a teensy-weensy bit obscure, but readers who have browsed in the history of chemistry and in mythology may have happened upon them. In 1600 Raphael Elgin named the supposed flammable essence of things *phlogiston*. In 1774, when Joseph Priestly discovered oxygen, he first named it dephlogisticated air. The coastal area at the foot of Mount Vesuvius in southern Italy was named by early Greek settlers "the flaming fields." Later Romans borrowed part of the Greek name and called it in Latin *Campi Phlegraei* "flaming country-side." Our English word *flame* is related to *phlox* through Latin *flamma*, earlier supposed form **flagma*. Another slightly hidden relative is a Greek medical term for inflammation, *phlegma* 'a burning sensation, an overwarming of body tissue,' which then came to refer to the mucus produced by certain inflammations, phlegm. In Greek mythology, the hot-headed son of Ares, the god of war, was fiery Phlegyas. "Fiery" was an esteemed warrior name in ancient Greece. One of the companions of Herakles (Hercules) in his battle with the Amazons was Phlogios.

∾

ROCK CRESS

Other common name: Wall Cress

Genera: Arabis < Greek adjective, of Arabia, probably because Rock Cress grows in dry places (?), but Arabis may also be a folk form for some now lost word of similar form. Highly fanciful suggestions have been made, for example, that Arabis is a hybrid of Greek and Latin, some unlikely form like **arabies* (*a* Greek, not + *rabies* Latin, raving) for some herbalist's notion that Rock Cress leaves cured rabies. They do not. Nor is there printed proof of such an origin.

Some species of a different genus, Aubrieta, are also called Purple Rock Cress, after a French artist who specialized in botanical subjects, Claude Aubriet (1668-1743).

Family: Cruciferae, the mustard family < *crux, crucis* Latin, cross + *ferre* Latin to carry, to bear. The cross they bear is the arrangement of their flower petals, usually four petals at right angles to one another, making the grouping resemble a little cross. Sweet Alyssum belongs to this family.

SPECIES
Arabis albida (Latin, whitish, of the flowers) provides a pleasant bed for some of the early spring bulbs, as it blooms when they do. In many Canadian gardens, Rock Cress blossoms at the same time as daffodils and primulas. A more exotic, interesting, and consequently difficult-to-find-at-a-Canadian-garden-centre

cress is *Arabis blepharophylla* (Botanical Latin, with leaves fringed with eyelashes < *blepharos* Greek, eyelid + *phyllon* Greek, leaf) which is native to California and will thrive in the Lower Mainland of British Columbia.

Purple Rock Cress, a Mediterranean native, is *Aubrieta deltoidea* (Botanical Latin from Greek, triangular, said of the petals). Deltoid means shaped like *d*, the Greek letter delta or *d* < *delta* + *eides* Greek, form, shape; hence -*oid* means 'formed like, similar to.'

WORD LORE

Cress looks as if it should be related to words based on *crescere* Latin, to grow, words like accrue, crescent, decrease, increase, and recruit. But it entered Old English from early Germanic *Kresse*, which is related to the Old High German verb *kreson*, to creep, to crawl.

∾

SILVER-DOLLAR PLANT

Other common names: Dollar Plant, Honesty, Lunaria, Money Plant, Moneywort, Moonwort (the old English name), Satin Flower

Genus: Lunaria < *luna* Latin, moon, from the flat, rounded, papery seed-vessel that suggests a full moon, and is part of many dried-flower arrangements. It also looks coin-like, hence Money Plant.

Family: Cruciferae, the mustard family < *crux, crucis* Latin, cross + *ferre* Latin to carry, to bear. The cross they bear is the arrangement of their flower petals, usually four petals at right angles to one another, making the grouping resemble a little cross. Sweet Alyssum and Rock Cress belong to this family.

SPECIES

Lunaria annua (Latin, annual) is technically a biennial but will produce its paper moons the first year, with the bonus of sweet-scented pink and white flowers in late spring or early summer. *Lunaria rediviva* (Latin, returning to life, that is, perennial) is a European relative.

∾

SPEEDWELL

Other common names: Veronica, Brooklime, Clump Speedwell, Woolly Speedwell

Genus: Veronica < supposedly named after Saint Veronica, but see *Word Lore* below.

Family: Scrophulariaceae, the figwort family, named after its typical genus, *Scrophularia auriculata* (Botanical Latin, with little ears, referring to the heart-shaped base of the leaves) which in turn was named because herbalists thought it would cure scrofula.

SPECIES

The old name for the traditional, European, blue-flowered perennial is *Veronica officinalis* (Botanical Latin, of the herbalist's shop < *officina* Medieval Latin, shop). The later and still current name is *Veronica spicata* (Latin, spiked, of the flower head). These and other species have long spikes of small florets that open from the bottom up through summer and early fall. Blue is the oldest colour, to which hybridists have added pinks, whites, violets, and art shades of blue.

WORD LORE

The popularity of Veronica as a female given name is due to Saint Veronica, a woman of Jerusalem whom pious legend states stood along the *via dolorosa* and took pity on Jesus and his suffering as he was led to his crucifixion. Veronica stooped down and wiped Christ's sweating brow with her veil. Later she noticed that the cloth now bore a true image of the face of Christ. At St. Peter's in Rome such a cloth, claimed to be the original veil of St. Veronica, may be seen among the holy relics in the Vatican collection. July 12 is St. Veronica's feast day, and girls born on this date are sometimes baptized with the name in the Russian Orthodox, Greek Orthodox, or Roman Catholic churches.

Like dozens of popular female names, it was early bestowed on a flower.

One standard but dubious etymology of Veronica arises from this religious story, and is the origin found in many, poorly researched "names-for-your-baby" books. Monkish folk etymology posits that the name is a contraction of the Latin phrase *vera iconica* meaning 'true image,' from the same Greek root that gives us the word for a religious painting, icon. Such a derivation is linguistically unlikely.

Much more probable is the derivation of Veronica from a medieval Latin adjective *veronicus* that meant 'a person from the Italian city of Verona.' Veronica would mean a woman of Verona, a suitable name for a girl because the women of that city were considered to be among the most beautiful in all of medieval Italy. Shakespeare places Romeo and Juliet "in fair Verona where we lay our scene." But Verona also claimed to possess the veil of St. Veronica at one point during the Renaissance! Sixteenth-century Italian referred to this as *la veletta veronica* 'the Veronese veil.' This, I believe, is the origin of the feminine first name. Véronique is the French spelling. In some Slavic tongues, it is Veronika or Beronika. The fact that one of the Catholic Stations of the Cross commemorates St. Veronica's kindness to the suffering Jesus insures the name's continued popularity in all languages of the western world.

Another, to me doubtful, source of the given name Veronica is the female first name Berenice (often shortened to Bernice), originally a Macedonian form of the Greek *Pherenike* 'bringer of victory.' The vowel shiftings and interlingual gradations necessary to support this bizarre transformation may happen on Mars, but have not so far occurred on earth, except in the frantic noggins of certain desperate etymologists.

Speedwell

A common name for Veronica since the sixteenth century, speedwell is obviously a verbal phrase, but what does it speed so well? Cures. Recommended by medieval herbal lore, an infusion of leaves and flowers was said to promote speedy recovery from all manner of mortal maladies. This may be related to that dubious etymology of Veronica that says the saint's name is based on a Greek form, Pherenike 'Victory-Bringer,' that is, the plant brings quick victory over disease.

∾

ST. JOHN'S WORT

Common names of related species: Aaron's Beard, Goldflower, Hypericum, St. Peter's Wort

Genus: Hypericum, Latin < *hypereikos* Greek, St. John's Wort. Dioscorides and Galen have a later form, *hyperikon*, from which Pliny borrowed the Latin version. The first Greek form appears to contain the Greek word for heath or heather, *ereike*, preceded by *hyper* Greek, above, over. Thus the prime meaning may be growing 'above or through the heath plant.' But the second root could also be *eikon* Greek, picture, image, from which current English derived *icon,* holy image of a saint or deity. This derivation was widely believed by early Christians, hence the habit of hanging a sprig of St. John's Wort above holy images, particularly above icons depicting saints on their feast days.

Yet another meaning of *eikon* produced yet another use of this plant. A common secondary meaning of *eikon* in ancient Greek was 'ghost, phastasm, ethereal spirit.' Hence a folk belief developed that *hyperikon* meant 'over a ghost' and so the herb would keep away "ghoulies and ghosties and things that go bump! in the night." Now the leaves of St. John's Wort have little oil glands that give them a dotted look. Folk superstition said that the devil was so angry about St. John's Wort driving off his evil minions that he pierced the leaves with these dots in a fit of satanic pique. From the immemorial past, Midsummer's Eve was a night when witches, goblins and volant sprites took flight and flapped on leathern wing or slithery broom to do nefarious deeds. It is the night of June 24, called *Walpurgisnacht* in German. Saints' icons were protected on this night by hanging St. John's Wort over them. This is also the origin of the English common name, for in the calendar of saints, the feast of St. John the Baptist is June 24, a date on which the plant was supposed to be in bloom. Another wisp of herbal folklore said that dew falling on the plant overnight could be collected on the morning of June 24 and rubbed on the eyes, which would safeguard one's peepers from disease.

Family: Guttiferae, the mangosteen family, named after a succulent tropical fruit of Malaya, *Garcinia mangostana*. Guttiferae < *gutta* Latin, drop, of the resinous juices and oils in tiny glands on the leaves of these largely tropical trees and shrubs + *ferre* Latin, to carry, to bear, to produce. Related to Latin *gutta* are gutter, a trough for water drops, and gout, a form of arthritis also called dropsy, once believed to be the release of drops of the bodily humours. The French-derived vocabulary of heraldry has terms like *guttée d'eau* 'water drop' and *guttée de larmes* 'tear drop' to name heraldic symbols appearing on crests and coats-of-arms.

Some botanists exclude St. John's Wort from this family and place it in its own family, Hypericaceae, a family of about eight genera and 350 species, more than 300 being *Hypericum* species. About 50 of these are native to North America.

SPECIES

Native to British Columbia and possible to use there as a pleasant groundcover is *Hypericum anagalloides* (Botanical Latin, like Scarlet Pimpernel or Anagallis, said of the tiny leaves). A noxious, introduced weed in much of southern Canada is *Hypericum perforatum* (Latin, with holes through it, said of the apparent holes in the surface of the leaf and at the edge of the petal, in fact these dots are the transparent oil glands). Light-coloured animals may be poisoned by some of the glucosides in the plant. The chemicals are phototropic, being turned by light into toxic nasties. Light-coloured animals and livestock are particularly susceptible.

The most common garden subject, also good for ground-covers in mild climates, is *Hypericum moserianum* hybridized by a botanist named Moser. Called Goldflower, it blooms in one of nature's prettiest yellow hues.

Uses

Elizabethan gardener and famous writer on plants, John Gerard, in his *The Herball, or General Historie of Plants* published in 1597 waxes positively ecstatic about the virtues of St. John's Wort. With modern caution in mind, we quote a smidgeon.

> "S. Johns wort with his floures and seed boyled and drunken, provoketh urine, and is right good against the stone in the bladder, and stoppeth the laske [diarrhea]. The leaves stamped are good to be layd upon burnings,

scaldings, and all wounds; and also for rotten and filthy ulcers. The leaves, floures, and seeds stamped, and put into a glasse with oyle Olive, and set in the hot Sunne for certain weeks together, and then strained from those herbs, and the like quantitie of new put in, and sunned in like manner, doth make an oyle of the colour of bloud, which is a most precious remedy for deepe wounds, and those that are thorow the body, for sinews that are prickt, or any wound made with a venomed weapon."

THRIFT

Other common name: Sea Pink

Genus: Armeria < a Renaissance Latinizing of a French name for a dianthus, *armoires*, so named because the dianthus had a typical thick flowerhead that vaguely resembled a chest used to store tools and small weapons.

Family: Plumbaginaceae, the leadwort family < *plumbago* Latin, leadwort < *plumbum* Latin, the metal lead, indicating that some plumbago species were thought to be remedies for lead poisoning. The Latin word for lead gives several current English terms. A plumber originally worked with lead pipes or plumbing. A plumb line is held taut by a lead weight at its end. Through *plomb* French, lead, comes a verb, to fall like heavy lead, that is, to plummet.

SPECIES
Armeria maritima is the common Sea Pink of gardens.

WORD LORE
Why is this plant called Thrift? One Elizabethan guess said it has many stalks coming from one root, and thus makes thrifty use of roots. That is exceedingly fanciful. The true origin of the common name is lost.

Other common name: True Valerian (Species in four other genera are sometimes called valerian incorrectly.)

VALERIAN

Genus: Valeriana, the medieval Latin name < *Valerius* Latin, the name of a Roman *gens*, a prolific family with many individuals mentioned in the literature, but who precisely the plant name honours is unknown. The female given name Valery is from the feminine form *Valeria*.

There was a late Roman emperor named Valerian, ruling from 253-260 A.D., who was an enthusiastic persecutor of Christians. Valerian received his comeuppance, however, when on a military campaign he was captured by the Persians in 260 A.D. Shapur the First, King of Persia, used the Roman emperor as a footstool for mounting his horse. After poor Valerian's spine gave out, Shapur had him killed and flayed. When Roman envoys later arrived to negotiate with the Persian king, he brought out the bundle of rolled skin and told the envoys, "This is what I do to Romans who annoy me." It seems unlikely the plant was named after that Valerian. However, if named for the Valerius family, note that their *gens* name is one of cheerful omen, being derived from *valere* Latin, to be healthy. For my personal guess at the origin of the name, see *Word Lore* below.

Valeriana hortensis.
Garden Valerian, or Setwall.

Family: Valerianaceae, the valerian family, of ten genera and about 400 species, mostly perennial flowers and shrubs, 200 of them in the *Valeriana* genus.

SPECIES

Valeriana officinalis (Botanical Latin, of the herbalist's shop < *officina* Medieval Latin, shop) was widely employed as a medieval anaphrodisiac (*ana* Greek, against + *aphrodisia* sexual feelings < *Aphrodite* Greek, goddess of physical love). Common Valerian is still grown in gardens for its fragrant pink, lavender, and white flowers.

Uses
In the middle ages, monks and nuns drank extract of valerian when they felt too frequent sexual urges. At least, some of them did. Oil of valerian is a sedative and a depressant acting on the central nervous system. Could this explain the lugubrious sonority of plainsong? Or the meek schnookiness of medieval nuns and the mild monkishness of their brethren? The chemical in valerian that produces sedation is a monoterpene called valepotriotes. Oil of valerian is still prescribed regularly by physicians in Europe as a remedy for hysteria (?), "nervous unrest," insomnia, and as an anticonvulsant and antispasmodic in epilepsy. Newer drugs with less chancey side effects have replaced valerian in the North American pharmacy.

WORD LORE
My second choice as the Valerian after whom the plant was named is Saint Valerianus, a bishop of Auxerre in the thirteenth century. Perhaps he first prescribed this fetid extract to quell unseemly passions coursing through the cloisters of Auxerre? Spoilsport. Yet I have unearthed one, final, more likely candidate in the lore surrounding Saint Cecilia, a Christian martyr of uncertain date. In her saint's tale, we learn that she was a Christian damsel of patrician family who chose to marry a pagan named Valerian. The charming Cecilia waited until her wedding day to inform her randy groom that she had been thinking things over—the premarital qualms of many a bride since—and she had decided to dedicate her virginity to God: "And well, Valerian, you won't mind if we don't actually engage in any major-league insertion, pudendumwise, will you, dear? Seriously though, Val, you do understand that all subsequent hanky-panky shall be of a refined and spiritual nature? Oh, I might occasionally let you buff my halo with the hem of your tunic, but, as a bride of God, I can no longer just

jump in the old cubiculum at the drop of a toga." The bridegroom Valerian agreed to respect her vow and converted to Christianity. Now, surely, this is the Valerian after whom a desexing potion might have been named. Valerian was afterward himself martyred and there is an historical record of that. Saint Cecilia was later brought before a Roman prefect as an obstinate Christian. She too refused to recant and was put to death.

∾

VIRGINIA BLUEBELLS

Other common name: Virginia cowslip

Genus: Mertensia < named after a professor of botany at the University of Bremen, Franz Carl Mertens (1764-1831)

Family: Boraginaceae, the borage family

SPECIES
Mertensia virginica, Virginia Bluebells, is the perennial wildling of eastern Canada and the United States. The fat little tubers are planted in the fall beside daffodils and early tulips, to blossom with flowers of a unique blue and provide a fine background for all white-flowered spring bulbs. But the genus is not confined to the eastern woodlands. In southern British Columbia, anyone wandering over sagebrush flats or climbing up hillsides spiked with Ponderosa Pine can enjoy, some late April morning, a turquoise carpet of *Mertensia longiflora* (Botanical Latin, with long flowers). At higher altitudes, Western Bluebells will bloom later at the end of May or early in June on much stubbier stems, but in the same rich Mertensian blue.

∾

WORMWOOD

Other common names: Dusty Miller, Lad's Love, Mugwort, Old Man, Old Woman, Silver Mound Artemisia

Genus: Artemisia, Greek name for one species < *Artemis* Greek, goddess of hunting and chastity, equivalent to the Roman Diana. There are several possible reasons for the name, none of them proven. First and in my opinion most likely,

Artemisios was the name of a spring month in ancient Sparta and in Macedonia, the time of the year when good hunting could once again be resumed after the rigor and scarcity of winter. The species of artemisia native to Greece bloomed as hunters returned to the woods and mountains, after a prayer to the goddess of the hunt, Artemis.

Second is an origin of the plant in the name of the mourning wife of King Mausolus of Caria. Her name was Artemisia and she so loved her dead husband that she superintended at Halicarnassus the building of an elaborate tomb for him, which became one of the seven wonders of the ancient world, known as the Mausoleum. This story was widely known in antiquity, quoted by Cicero and Pliny, from older Greek sources. Some species of artemisia look distinctly droopy and mournful, particularly in bloom. Most species have a silvery-whitish coating on the leaves. It may be recalled that a frequent token of mourning in many ancient Mediterranean countries was for members of the funeral party to pat flour on their faces and not remove it until the burial or cremation had occurred. Could such people have seen in these old Greek "Dusty Miller" species, a sign of floury mourning ritual?

Family: Compositae, the daisy family < *compositus* Latin, placed together. This largest family of flowering plants is named after the compound flowers of its members. Small florets of individual flowers make up large clusters or heads.

SPECIES

Tarragon
That most essential herb of French cookery, known as French tarragon, is *Artemisia dracunculus*. The specific means 'little dragon,' a reference to the complicated etymology of tarragon. It seems to be something like this: *drakon* Greek, dragon > *tarkhon* Arabic, dragon and this herb > *tarkhon* Medieval Greek, the word borrowed back into Greek > *tarchon* Old French > *estragon* Modern French > *tarragon* Modern English. Fascinating, but why is the plant associated with dragons? Did ancient herbal folklore think a sprig of tarragon would fend off those nasty fire-breathers? Perhaps.

Absinthe, its manufacture now banned, was a green French liqueur flavoured with oil of wormwood, *Artemisia absinthium* (Botanical Latin from *apsinthion* Greek, ultimate origin unknown). One of the foliage plants called Dusty Miller is *Artemisia stelleriana* named after Georg Wilhelm Steller (1709-1746), a German collector of botanical and zoological specimens who discovered the species in Siberia. One of the plants called Sagebrush, pesky weed of the West, used to be called *Artemisia tridentata* (Botanical Latin, three-toothed, said of the leaves), but has now been reclassified and given its own little genus with a name based on *seriphon*, one of the Greek wormwoods. Sagebrush is now officially dubbed *Seriphidium tridentatum*. *Artemisia canadensis* or Wild Wormwood is infrequently a garden subject. Its leaves have no hairs and it bears multiple and curious green flowers.

WORD LORE

Mugwort was *mycge-wyrt* in Old English, that is, a vermifuge, a plant that keeps away midges, tiny gnat-like insects. This yellow-flowered *Artemisia vulgaris* (Latin, common) is native to Eurasia but has been naturalized in North America where it is often a troublesome weed.

∾

Other common names: Milfoil, Sneezewort

YARROW

Genus: Achillea < *Achilles* Greek, hero of Homer's *Iliad*. The great warrior was taught the use of yarrow as a vulnerary, a healing herb whose leaves and flowers were made into a salve applied to battle wounds by the ancient Greeks.

Family: Compositae, the daisy family < *compositus* Latin, placed together. This largest family of flowering plants is named after the compound flowers of its members. Small florets of individual flowers make up large clusters or heads.

Millefolium luteum.
Yellow Yarrow.

SPECIES

The dried roots of Sneezewort (*wyrt* Old English, plant) or *Achillea ptarmica* have been used for thousands of years to make snuff. The species name is from *ptarmike* Greek, the sneezing herb < *ptarmos* Greek, a sneeze. Common yarrow with its finely divided, fern-like leaves is *Achillea millefolium* (Botanical Latin, with a thousand leaves). As *gearwe,* Old English borrowed the plant name *yarrow* from Scots Gaelic, where it means 'rough stream,' and is the name of Scotland's Yarrow River, as well as a place in the county of Selkirkshire. Growing throughout the Northern Hemisphere, yarrow is native to Canada, as is a particularly woolly subspecies, *Achillea lanulosa* (Botanical Latin, very woolly).

Note that the Elizabethan genus name has become the modern specific name for common Yarrow.

TREES

NATIVE TO CANADA

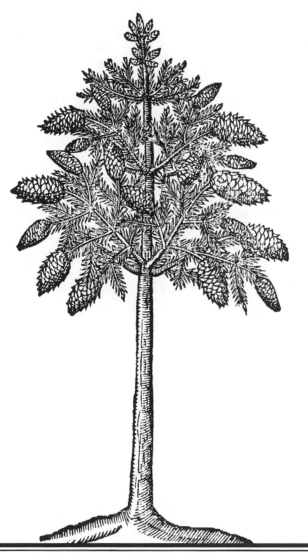

TREES NATIVE TO CANADA

Genus: *Alnus*

ALDER

Family: Betulaceae, the birch family < *betula* Latin, birch

French: *l'aune*, older spelling *l'aulne* < *alnus* Latin, alder

WORD LORE

Alder < *aler, alor* Old English. The letter *d* inserted in dialect pronunciations was fixed in manuscript English by the fourteenth century. The Latin, French, and English words for alder tree all hark back to a common Indo-European root in *al and *el. The people who spoke Proto-Indo-European knew some betulaceous tree like the alder. Their root appears in languages like Old Scandinavian (what Vikings spoke) *elrir*, *ölr*, in Old High German *elira*, in Modern German *Erle*, in Lithuanian *alksnis*, and in Polish *olcha*.

Notes

What do dam-building Canadian beavers and sixteenth-century Venice have in common? Underwater pilings made of alder wood! Beavers (*Castor canadensis*) build dams with underwater portions often composed of alder boughs which are superbly waterproof once submerged. Beavers also use and eat aspen and birch. Venice, with its Latin motto, *Serenissima Respublica*, Most Serene Republic, is said to be held above the encroaching waves on once stout pilings made of a European species of alder, probably *Alnus glutinosa*, the Black Alder, in Canada an imported species planted as an ornamental tree. The Black Alder's young leaves are glutinose, that is, clammy, so sticky in fact that some medieval Europeans scattered alder leaves on the floors of their houses to trap fleas. In medieval France and Holland, waterproof alder wood was widely used to build sluices and troughs. The dyer's craft sees alder bark and young shoots used to make black, red, and yellow dyes.

Alnus hirfuta.
Rough leaued Alder.

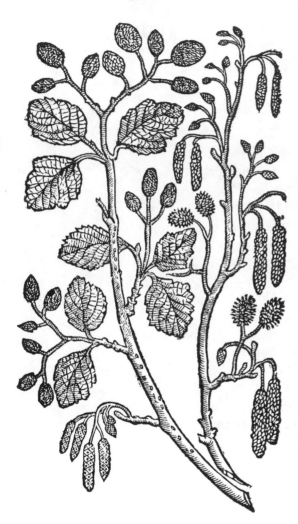

Canadian place names

Humans and alders share one habitat preference. Both like to live along streams. Hence the large number of Canadian, British, and American place names that include the *-alder* root or an earlier version of it. The village of Aldershot, now part of Burlington, Ontario, was named after a British military base. Its name in Old English was *Alre-sceat* 'alder-shoot.' A *sceat* was a piece of land that projected or shot out into a lake or other body of water. At the town of Aldershot north of Kentville in Nova Scotia, Canadian army recruits trained at Camp Aldershot before and during World War I and II. The Ontario and Nova Scotian towns were both named after an army base in Hampshire, England, made famous after it was founded to train British troops sent to fight in the Crimean War between 1853 and 1856.

Other Alder Places

Alder Point, north of Sydney Mines on Cape Breton Island, Nova Scotia

Alderford (narrow crossing place on a river where alders grow), England

Aldergrove, British Columbia, near Abbotsford

Alderholt (*holt*, Old English 'woods, thicket') England

Aldridge (in 1086 A.D., in the Domesday Book, it was listed as *Alrewic* 'alder dwelling' or 'farm')

The Aulneau peninsula (see *French Surnames* below) juts into Lake-of-the-Woods in northwestern Ontario.

Ruisseau aux Aulnes (Alder Creek) empties into the Etchemin River southeast of Québec City.

SURNAMES

English Surnames

Alder is a surname in Northumberland. Some instances of the surname Aldridge are from two Old English roots that mean 'farm beside alder trees' or 'ridge with alders.'

French Surnames

Aulne and Aune are both French surnames derived from an ancestor living beside alders. So too are French last names like Launay and Delaunay, as well as Aunaye, and the diminutive forms: Aunet, Alnette, Aulneau, Auneau, and Aunillon.

Other Surnames

Latvian has a surname, Alksnis 'alder.' Russian has Ólchin or Vólchin 'alder.' To step outside the Indo-European family, one finds a Finnish last name like Leppanen (or Leppainen) 'alder.'

German

Alder tree in German is *die Erle*. The German surname Erler recalls a founding ancestor's home in a grove of alders. The Erlkönig or Erlking was a nasty Teutonic tree goblin who lurked in dark forests. The spiteful sprite's name means 'King of the Alder,' his preferred abode. The Erlking lured children to their doom, as in the poem by Goethe, set to music by Loewe, and memorably by Schubert. "*Wer reitet so spät durch Nacht und Wind?*" Who rides so late through the night and the wind? It is a father on horseback with his son. The boy complains that the Erlking is choking him. But the father can't see the goblin, and, at the end of this gruesome *Lied*, the child ends up dead. Charming little ditty for an early nineteenth century evening of song in the parlour!

SPECIES

Five of the world's thirty alder species are native to Canada. Most are shrub-like rather than tree size.

Speckled Alder, widespread across Canada, is extremely important to its micro-ecosystem. Nitrogen-fixing bacteria on its root nodules make alder one of the few trees in the world that can fix nitrogen in a form plants can take in. The bacteria change gaseous nitrogen in the air into ammonia. Their high nitrogen content makes alder leaves decay quickly providing aquatic and soil nitrogen, especially useful when the tree colonizes new areas after a forest fire or after clear-cutting by lumber companies. Early-flowering alder catkins are magnets for bees. Alder seeds and flowers feed golden-winged warblers, hummingbirds, flycatchers, woodcocks, pine siskins, redpolls, and many other Canadian birds.

The Sitka Alder of British Columbia, usually shrubby, can be planted to ease soil erosion and control stream flow.

Quotation
There was in English a now obsolete adjective, aldern, formed like oaken. Here is a recipe not to try from *The Historie of Foure-footed Beastes*, a bestiary by Edward Topsell printed in 1607, in its original spelling: "If the right eye of a Hedge-hog be fryed with the oil of Alderne or Linseed, and put in a vessel of red brasse, and afterward anoint his eyes therewith, as with an eye-salve, he shall see as well in the dark as in the light."

APPLE

Genus: Malus
malus Latin, apple tree < *malum* Latin, apple < *melon* Greek, apple, or any tree fruit < Indo-European root **mel* sweet-tasting, like honey or ripe fruit

Family: Rosaceae, the huge and perhaps too encompassing rose family

WORD LORE
Apple < *aeppel*, *aepl* Old English. The word *apple* is related to Modern German *Apfel*, Dutch *appel*, Swedish *äpple*, and Danish *aeble*. Slavic languages add a diminutive to the root to give Russian *yablo-ko*, Polish and Czech *jablko*, and Serb *jabuka*. Related forms appear in Welsh *afal*, Gaelic *ubhal*, Lithuanian *obuolas*.

In *Origins*, Eric Partridge presents a putative origin for this European *apple* word, from the name of a Roman town noted for its apples and other fruits and nuts, Abella, an ancient marketing centre for produce from the surrounding fields of

fertile Campania. The Latin poet Virgil called the little town
Abella malifera 'Abella rich in apples.' Today it's Avella, a few
kilometres inland and east of Naples. How important was this
place for fruits and nuts? Well, the Italian word for hazelnut is
avellana 'of Abella,' and the botanical name for the hazelnut
tree is *Corylus avellana*. A rare synonym in English for hazelnut or filbert is avellan. In heraldry there is a design called the
avellan cross, a stubby crucifix shaped like four stylized hazelnuts.

As for Partridge's contention that the word *apple* stems
from the name of the town, it is much more likely that Abella's
municipal name is from an Indo-European root **abel* meaning
'fruit of any tree.'

Canadian Apples

From some of the roughly two dozen true
species native to the northern temperate zone,
including the crab apples, have come the
thousands of hybrids and sports that produce
modern eating apples. The process took centuries of selection, grafting, and hybridizing
of naturally small-fruited wild species like
Malus sylvestris (Latin, of the forest), *Malus
pumila* (*pumilus* Latin adj., dwarf), and *Malus
prunifolia* (Latin, with leaves like a plum
tree). Apple trees shipped from France grew at
Annapolis Royal by 1635. Our earliest
settlers used apples principally to make cider.

The McIntosh Red is Canada's most
famous apple. Farmer and apple-breeder John
McIntosh immigrated from the Mohawk
Valley to Iroquois in Upper Canada in 1796.
By 1811 he was clearing land at nearby
Dundela when he discovered an old orchard.
One of the twenty trees bore very tasty apples.
His son Allan McIntosh grafted stock of the
original tree in 1835 and went into the apple
business big-time. The McIntosh trees tolerat-
ed varied soils and climates. Near the site of
the original tree, that died in 1910, is an his-

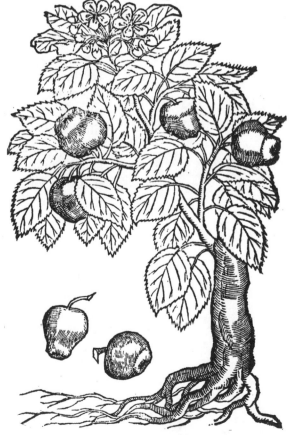

Malus Carbonaria longo fructu.
The Bakers ditch Apple tree.

torical plaque that says the original McIntosh tree bore apples for ninety years! Once, in England, I spoke of eating a McIntosh. My listener, a tobacconist in Leeds, stared at me as though rising damp had risen too far up my brainstem. In the British Isles, a mackintosh, a mac, or a mack, is a rainproof coat, invented by Scottish scientist Charles Macintosh by laminating two layers of cloth with rubber.

The *Fameuse* apple, a cultivar based on French stock, is still grown sporadically in its native Québec. Other once-popular Québec apples were the *Pomme-de-neige* 'snow apple,' *Pomme gris(e)* 'grey apple,' the *Bourassa*, and the St. Lawrence. *Pomme-de-glace* 'ice apple' was an early Acadian cultivar.

Two crab apples are native to Canada. The Pacific Crab Apple grows only in British Columbia. The hard, heavy wood of the crab apple makes it of little commercial importance. But pioneer grist mills in the Ontario countryside had gears and blades of crab apple wood.

Canadian Paring Bee

Early settlers in the Canadas (Upper and Lower) liked to combine socializing with work. One such neighbourly gathering was the paring bee, also called apple frolic, paring frolic, and apple bee. It served the social function of preliminary courting ritual where young men and women from nearby farms could scout potential mates, while older married couples could gossip and exchange local news. Apples were pared, sliced, and hung to dry on long strings that might be suspended from rafters in the kitchen, attic, or cold cellar. A staple at Canadian pioneer tables throughout the winter months was stewed dried apples.

Canadian Apple Songs

"In the shade of the old apple tree." Canadian lyricist Harry Williams wrote the words to this Tin Pan Alley hit song of 1905. In his *Canadian Quotations*, editor John Robert Colombo reports that the lyrics were inspired by one actual apple tree that grew on Glen Edith Drive in Toronto. Williams was also lyricist of "It's a Long Way to Tipperary," adopted as the marching song of the British army in 1914. In the 1970s, George Hamilton IV wrote a country ballad, "When It's Apple-Blossom Time in Annapolis Valley."

Canadian Place Names

Apple Hill, near the town of Casselman in southeastern Ontario

Apple River, Nova Scotia

Apple Tree Landing, former name of Canning, Nova Scotia in the Annapolis Valley. Annapolis Valley apples enjoyed a long and bountiful export trade to Britain until the late 1920s.

SURNAMES

English Surnames

Apple trees were a method of identifying fields and houses of founding ancestors, and thus contributed to several English surnames:

Apperley	'clearing with apple trees'
Applin	'(among) the apples' from the Old English dative plural *aepplum*
Appleby	'apple farm' with Old English *by* 'a farmhouse' then 'a village,' with cognates in Swedish and Danish *by*, all akin to Viking word *byr* 'a farm'
Appleford	from one of several place names describing the shallows of a river where livestock could cross easily and where apple trees grew beside the ford
Applegarth	'apple enclosure'
Applegate	ultimately from Old Scandinavian *apaldrsgarthr* 'apple-orchard'
Appleton	'apple farm' with Old English *tun*, a common suffix on English place names. The meaning of *tun* expanded through history. Its initial sense was a hedge, a fence, an enclosure, a homestead, then a farm, a manor, a settlement,

a hamlet, and finally a village. Our modern form of the word is *town*.

Applewhite   from Applethwaite 'a clearing with apple trees' where thwaite is Old Norse for 'meadow, enclosed land'

Appleyard   a place in the West Riding of Yorkshire and a country synonym for orchard. Old English *geard* 'enclosure' (pronounced *yard*) is related to garden, and an Old Scandinavian cognate *garthr* gives English "garth," an open space within a cloister, yard, garden, or paddock.

Jewish Surnames

Applebaum is a partial Englishing of the Jewish surname Apfelbaum 'apple tree.' Variant spellings may indicate the origin of the founder. Among Polish, Hungarian, and Russian Jews, those speaking western Yiddish, it can be Apelboim or Appelboym. Eastern Yiddish speakers, like some Lithuanian Jewish families, may spell it as Apelbeym.

In certain German Jewish families, their surname is traceable directly to the medieval Frankfurt ghetto called the *Judengasse* 'The Street of Jews.' House numbers and street names were not common until the end of the eighteenth century, simply because most ordinary Europeans were illiterate. Houses and shops bore signs to identify their owners. Among the tree signs identifying houses and owners, and then developing into Jewish and German surnames, were Apfelbaum 'apple tree,' Birnbaum 'pear tree,' Buxbaum 'box tree,' Grünbaum 'green tree,' and Nussbaum 'nut tree.' These house signs sometimes also served to identify a shop or place of business. So an Apfelbaum might have sold fruit.

Some Other Surnames

Several Russian last names contain *jabloko* 'apple,' for example: Jablokov, Jablockov, Jablockin. Ábele 'apple tree' is a surname in Latvia. Armenian has a surname derived from an ancestor's nickname, occupation, or from living beside an orchard. This name is sometimes transliterated as Chendzorian with the familiar Armenian patronymic suffix *ian* prefixed with the Armenian word for apple.

The Golden Apple of Discord in Greek Mythology
According to Greek mythology, a bitchy goddess and an apple caused the Trojan War. Eris—her very name meant 'strife' in ancient Greek—was the evil bringer of discord, invoker of nasty spats during din-din, of squabbles before nighty-night sex, and so forth. Naturally, Zeus and the other gods didn't care to invite Eris to cloud parties, fly-by orgies, unveilings of statues of Zeus, or Olympian brunches. Eris was a bitch and she paid the other gods back with nasty tricks. One night a big wedding feast had excluded Eris, but invited three other goddesses: Aphrodite, Hera, and Athena. So Eris tossed a golden apple into the feast hall. On the apple were the words: "For the most beautiful female of all." The wedding guests asked Zeus to judge which of the goddesses should win the golden apple. No way, thought Zeus wisely, especially as one of the contestants, Hera, was his wife. If Hera didn't win—hooooo-boy! And the boy would be Zeus's favourite little cupbearer, Ganymede. But still, Zeus would never Hera the end of it.

So Zeus fobbed off the choice on a local princeling, Paris. Paris was the dim son of the king of Troy. Just then Paris was doing the shepherd thing in an adjacent ferny dell with a moist nymph, when suddenly by magic the three contending goddesses appeared, strongly suggested interrupting the coitus, and getting on with the judging. Now each of the goddesses was of a certain age and had no illusions about their beauty, but still, divinity has its perks. Also they knew Paris was a weakling, a coward, and chiefly interested in the old *ana-kata* (Greek 'up-down') atop nubile wenches. Therefore the three goddesses offered him bribes. Hera promised to make Paris lord of the known world. "Yeah, Yer Altitude, but would that include dancing girls?" Athena said she'd help Paris, a Trojan, win the Trojan War and knock those Greeks back into their giant olive jars. "And maybe throw in a couple of sacred virgins?" But the goddess of love, Aphrodite, had Paris's number. She offered Paris the fairest woman known to earth. The bad "judgment of Paris" awarded the golden apple to Aphrodite. At the time, the most beautiful woman in the world was Helen, already married to the Spartan Menelaus. Paris nabbed Helen anyway. And the Trojan War broke out because all of Menelaus's Greek allies pledged to help get Helen back.

Biblical Apples

The apple and apple tree are ancient symbols, mentioned in the Torah, Old Testament, and other ancient religious texts. Now this is not the fruit in Genesis, not the apple of temptation in the garden of Eden, a piece of which, folklore says, lodged in Adam's throat to give us the English phrase *Adam's apple*, referring to the little swelling in the front of the neck caused by the projection of the thyroid cartilage of the larynx. No, the pious took note of the apple and apple tree in Proverbs and, for example, in the Song of Solomon 2:3-5: "As the apple tree among the trees of the world, so is my beloved among the sons. I sat down under his shadow with great delight, and his fruit was sweet to my taste. … Stay me with flagons, comfort me with apples, for I am sick with love."

Red-faced generations of skittish explicators have rushed to convince readers that the Song of Solomon is not what it plainly is: a luscious passage of ancient Hebrew erotic poetry, composed to be chanted at a wedding feast as a celebration of sexual desire and love. Some commentators have insisted that the apple tree here symbolizes the love between God and his people, Israel. Well, read the Song of Solomon for yourself. The apple tree is a symbol alright, a symbol of a big, healthy, potent bridegroom, full of seed and ready for his wedding night. Later Christian writers also blushed to find such piquant sensuality enshrined in Holy Writ, so Roman Catholic dogma states it's all about the love between Christ and the Church. Not to be outpurged by mere papists, Protestant divines swoon in the deeps of the Song of Solomon as well, assuring all sex-hating sects that the poem concerns the love between God and man's soul. Not in the apple of my eye.

Biblical apples harbour a few other worms of contention. The Biblical Hebrew word pronounced 'tapPU'ach' in modern Hebrew and now meaning 'apple,' could never have referred to the fruit of the genus *Malus* that we know today, made big and juicy by hybridizers only in the last two hundred years. No species of the *Malus* genus grew in the hot places of the ancient Middle East. Now listen to how the Bible describes its 'apple': Joel 1:12 says it was a tree of the field like the vine, fig, and pomegranate. The Song of Solomon 2:3 and 7:8 says the apple had a sweet perfume and taste. Other passages say it hung in a tree that offered much shade.

I don't want to upset your apple-cart but Biblical scholars believe the fruit referred to by the word *tappuach* was an orange, a quince, or, most likely, an apricot. Ancient Palestinian folk wisdom said the apricot possessed aphrodisiac qualities, so its use as a sexual metaphor in the Song of Solomon is most apt. And apricot trees grew in ancient Palestine.

The Pome Words

French borrowed not the Latin word for apple, *malum*, but a word that meant fruit of any kind, *pomum*. This gave the French *pomme*, and several English terms borrowed from French. Pome is a botanical name for a kind of fruit like an apple or pear, with a thick, fleshy outer layer, and an inner core of seed capsules. James Joyce punned on the word in the title of a little collection of his poems, *Pomes Penyeach* (1927).

A perfumed ointment for dressing hair, originally containing decocted apple mush (Yech!), was a pomade. Medieval French had *pome d'ambre* 'apple of amber,' a ball of pleasantly scented substances in a mesh bag used to keep stored cloths and linens smelling fresh, and, in times of plague, foolishly carried on the person to ward off infection. This went into Middle English as a pomander, which could also be an orange or apple studded with cloves.

In cider and apple-juice making, the pulpy remnant left over after the pressing of the fruit is pomace, used to make commercial pectin or to feed cattle and pigs.

That torture instrument of the high-school gym class, the pommel horse, has an apple behind its name. A pommel, from an assumed diminutive form *pomellum* 'little apple,' came to mean the knob on the hilt of a sword, the knob at the front of a saddle, and then the leather-covered handles on the top of a pommel horse.

Pomology is the science of growing fruit trees. The Roman goddess of fruit trees was Pomona who gave her name to American towns in California, Missouri, and North Carolina, among other states.

Quotation

Roman orchardists made offerings to Pomona each spring to ensure the fertility of their fruit trees. A trace of this folk rite may linger in the old English custom of wassailing orchards on Christmas Eve. Hot cakes were put in the boughs of the best bearing trees, and warm cider was sprinkled on the bare branches. Then the apple or other trees were toasted. Here is one of the toasts as reported in Mrs. Grieve's *Modern Herbal* (1931):

> Here's to thee, old apple tree!
> Whence thou may'st bud, and whence thou may'st blow,
> Hats full! Caps full!
> Bushel—bushel-bags full!
> And my pockets full too! Huzza!

But perhaps this little carol of peasant greed offended the apple trees? After all, they had good taste. I like to think a few of the more refined trees pelted the wassailing twits with wizened pippins, much like the sportive arboreals that Dorothy encountered on the yellow brick road to Oz.

ASH

Genus: Fraxinus. Linnaeus named the genus from *fraxinus* Latin, ash tree

Family: Oleaceae, the olive tree family

French: *frêne* < *fraxin(us)* Latin, ash tree

WORD LORE

Ash < *aesc* Old English, akin to roots in widespread languages: German *Esche*, Old Norse *askr*, and Armenian *haçi*. This may be a tree name that predates even Indo-European. Egyptian hieroglyphics have a word for a kind of cedar tree, transliterated *ash*!

A letter of the Old English alphabet was called *aesc* or ash. It was æ, a ligature or duo of joined letters, now represented as a digraph, ae, representing a vowel sound midway between *a* and *e*. *Aesc* was one of the letters of the earliest Germanic alphabet called runes, an adaptation of Roman and Greek letters, modified to make carving them on wood or stone easier. An ash-plant is a walking cane, a stick, or a whip handle made of this wood.

SPECIES

Ancient Romans made javelins from the European ash, *Fraxinus excelsior* (Latin, taller, comparative of adj. *excelsus*). That ash is planted as an ornamental in Canada. North American aboriginal peoples made snowshoe frames, fish spears, and canoe ribs from strong but pliable white ash, *Fraxinus americana*. Black ash bark wrapped wigwams. Canadian hockey sticks—5 million of them a year—are made from white ash. Some outdoorsy types crush white ash leaves and rub the juice on mosquito bites.

Of the sixty world species of ash, four are native to Canada. The rarest is Blue Ash, *Fraxinus quadrangulata* (Latin, with four angles), found only in Ontario at Point Pelee, Pelee Island, and in the Thames River valley. The common name is from a blue dye obtained by chopping up the bark, soaking the pieces in water for a day, and boiling to concentrate it.

Some First Peoples wove baskets of Black Ash, *Fraxinus nigra* (Latin, black) after soaking and then beating it with stones during which Black Ash peels along its growth rings into thin slats.

MYTHS

1. Some native North Americans believed ash trees drove away snakes. Moccasins were sometimes lined with white ash leaves to prevent snake bite.

2. Glooscap was a legendary warrior and magician among the Maliseet, Mi'kmaq, and Abenaki peoples of eastern Canada. One Algonkian legend about the origin of humans tells of how Glooscap fashioned a mighty arrow of hard ash and then shot it into the great World Ash, the primordial tree, and watched delighted as the first human beings emerged from the wound in the bark.

3. The ash as World Tree, guardian of creation, hangs over Norse mythology as well. According to ancient Scandinavian myth, this tree was called Yggdrasill, and in Old Norse sagas it was labelled *askr* 'ash tree.' This ash linked the different worlds of the old Viking cosmos. Under its roots lurked the dour frost-giants, the realm of the dead, and deeper still at the underground spring of Fate was the Norse gods' daily council place where they convened to settle problems. Coiled round the ash trunk was a giant serpent who gnawed at the tree. The gods created women and men from trees growing near Yggdrasill. The ash linked the worlds of the gods, of the humans, of the giants, and of the dead. The World Ash was a source of renewal and nourishment. Unborn souls awaited life in its branches.

Many northern peoples had sacred woods where shamans performed rites beneath the boughs of magic trees. A later, related belief concerns some myths of the crucifix, which replaces the World Tree in Christian mythology. One ancient story says the cross on which Jesus was crucified when it was raised at Calvary stood at the centre of the known world on the very spot once occupied by the tree in Eden from which the serpent bade Eve pluck the fatal fruit of expulsion.

A violent echo of the Christian story was told in which the chief Germanic god Wotan crucifies himself to the World Ash Tree to obtain knowledge of the secret runes. In other words, the gift of runic letters, of literacy, was worth sacrifice, even by a god.

4. The ancient Greek word for ash tree was *melia*. There the ash was associated with magic rain-making rituals and thunderstorms. The ash 'called down the fire of heaven,' that is, it was prone to be struck by lightning. Trees struck by lightning were one source of fire for early humans. Greek mythology also

included the *Meliai* 'ash-nymphs,' amiable wee deities who
could be invoked to assist in rain-making. They also hovered
in beneficent attendance during the third month of the ancient
Greek sacred year, lambing month, whose symbol was the ash
tree. One of the Meliae (the Latin and English spelling) was a
foster-nurse of Zeus, again an association with a life-giving
liquid, milk here, rather than rain.

SURNAMES

Surnames in Fiction

Ashley Wilkes is the noble goody-two-shoes, aristocratic and
slender as an ashen wand, for whom Scarlett pines in *Gone with
the Wind*. Author Margaret Mitchell found the character's first
name common among southern gentry, the Christian name
based on a British surname, itself stemming from an ancestor
who lived at a lea (forest clearing) among ash trees.

Gustav Aschenbach (German 'at the stream where ash trees
grow') is the German academic who vacations before his *Death
in Venice* in Thomas Mann's famous novella, suffering a fatal
infatuation with a beautiful Polish boy.

Well-known Ash Surnames

Dame Peggy Ashcroft (1907-1991) was a renowned
Shakespearean actress who won an Academy Award for her
luminous performance as the British mother in David Lean's
film version of E. M. Forster's novel, *A Passage to India*
(1984). Her family name comes from the place in the English
county of Berkshire. A croft was arable enclosed land next to a
house, in Ashcroft, such a field with ash trees.

Arthur Robert Ashe (1943-1993) was the first black
American male tennis player to win the U.S. Open and
Wimbledon.

Laura Ashley (1925-1985) was a Welsh designer of textiles,
wallpapers, and clothes who revived Victorian and Edwardian
floral patterns in the 1960s to create by the mid-1970s a worldwide business.

Hal Ashby (Old English from Danish 'settlement near an
ash grove') is a prominent Hollywood film director.

Maurits Cornelis Escher (1898-1972) was a Dutch graphic artist of great popularity and some drafting skill. His name is Germanic from an ancestor who dwelt beside an ash tree (*die Esche*, German, ash tree).

French Surnames from *frêne*

Ancestors who lived beside ash trees gave rise to French surnames like that of Pierre Fresnay, stage name of a French film star active in movies from the 1930s to 1950s. Fresnay is a common last name and village name in France, from Old French *fraisnaie*, modern French *frênaie* 'ash grove.'

Philip Morin Freneau, poet of the American Revolution, had a French surname that meant 'little ash.'

The Fresnel compound lens, once used in lighthouses and still employed as a diffusing and effects lens in commercial photography and film work, was invented by French physicist Augustin Jean Fresnel, whose surname goes back to a medieval Latin diminutive, *fraxinellum* 'little ash.'

Trent Frayne, a Canadian journalist, has a Norman French surname that means an ancestor dwelt by an ash. Many variations of Old French *fresne* occur in England and France, among them: Frane, Frayn, Frayne, Freen, and de Freyne.

Spanish Place Name

The city of Fresno, California, took its name from a local river dubbed by Spanish explorers for a nearby ash grove. The Spanish word for ash tree is *fresno* < *fraxinus*, Latin.

Quotation

In the Department of the Interior Annual Report for 1877, an entrepreneurial Canadian bureaucrat reported on wearing apparel he had noted among the Wabanakiyak people of eastern Canada: "I think ... the ash hats, for men and women, manufactured by the Abenakis, will rival in value and quality the Leghorns and Panamas [hats also made of natural fibres]."

❧

Genus: Populus, named by Linnaeus < *populus* Latin, poplar. This root is *not* related to the 'people' words in Latin like *populus, publicus, plebs*. It is from a distinct Indo-European tree word that has nothing to do with the Latin source of English words like *population* and *depopulate*. See also *Poplar* entry.

Family: Salicaccae, the willow family < *salix* Latin, willow tree

WORD LORE

Aspen < *aesp, aspe* Old English. In Middle English the simple noun form became obsolete and was replaced by aspen, the adjectival form. The Indo-European root * *apsa* 'poplar' is not related to the root in asp, a snake, but it does evolve into German *Espe* and holds firm in Latvian *apsa*.

French: *tremble* < *tremulus* Latin, the 'quivering' tree *peuplier* < *populus* Latin, poplar, aspen; also called *le peuplier faux-tremble*, false-*tremble* poplar, *Populus tremuloides*, Eng. trembling aspen

SPECIES

Aspen is found everywhere in southern Canada and north to the tree line. Six of the thirty-five world species are native to Canada, including the most familiar one, *Populus tremuloides*, trembling or quaking aspen, whose leaves shimmy in even the faintest current of air. Aspen leaves do their familiar

ASPEN

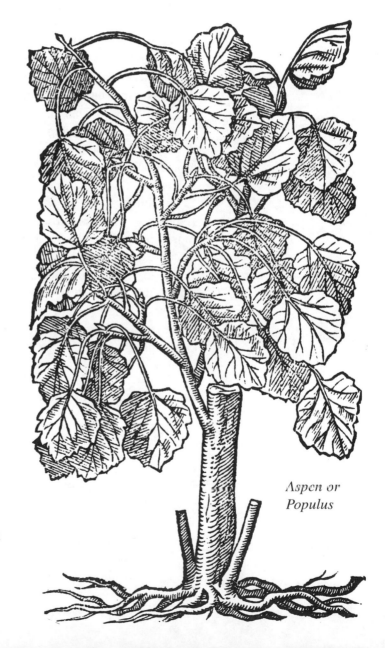

Aspen or Populus

dance because their petioles (leaf stems) are longer and flatter than those of most other trees. A European folk story explains the quaking leaves another way. The aspen leaf trembles from shame and horror because the cross on which Jesus Christ was crucified was made from aspen wood. The fact that one species, *Populus tremula*, did grow in the ancient Near East helped spread this pious piffle.

By tradition, lumbermen sneer at aspen as a junk tree, suitable only for making pulp. In nature, however, the only true "weeds" in a forest are bozos with power saws. The quick growth and invasive spread of aspen after a fire or clear-cutting stores energy in forms other woodland creatures can use, like the beaver in winter whose chief chomp is aspen bark.

Another Canadian native is Largetooth Aspen, *Populus grandidentata* (Latin, big-toothed), identified in spring by soft down on buds and twigs, in summer by leaves with large-toothed margins.

Place Name

Aspen, Colorado, is a resort of arriviste yahoos often busy murdering one another with Magnum .357s or unveiling to an anxious world independent film treasures like "Poodles in Bondage." Here ski-bunnies and ski-bums seek status as they slalom into the *haute bourgeoisie*. The process is complex, but involves delicate rites of passage such as paying $189 U.S. at a restaurant for fondue for two: "Youse want **cheese** wid it? Dat's extra, man."

SURNAME

Russian has Osinin < *osina* 'aspen.'

Quotation

Alfred, Lord Tennyson, that melodious technician of Victorian sentiment, found this tree word useful in *The Lady of Shalott*:

> Willows whiten, aspens quiver,
> Little breezes dusk and shiver
> Thro' the wave that runs for ever
> By the island in the river
> Flowing down to Camelot.

Other common names: Linden, Lime, Bee Tree

BASSWOOD

Genus: Tilia < *tilia*, Latin, lime or linden tree. The Latin name for the European linden is probably akin to *tillein*, Greek 'to pluck [hair or feathers], to extract fibre from.'

Family: Tiliaceae, the linden family

French: *tilleul* < *tiliolus*, medieval Latin diminutive form 'little linden tree.' In France linden-tree flowers were dried and made into a calming tea, *une tasse de tilleul.*

WORD LORE

Basswood

Bast < *baest* Old English, the root common to many Teutonic languages, like modern German *Bast*, Dutch *bast*, Old Norse *bast*, akin to Latin *fascia*, literally 'bundle bound by fibrous rope.' Basswood goes back to *bass*, a British dialect pronunciation of bast. Originally bast was fibre from the phloem or vascular tissue in the inner bark of the linden tree, of ancient and widespread use throughout Europe and among North American First Peoples. The long-celled bast fibres were made into rope, into twine, into plaits for basket weaving, mats, etc. The word *bast* was later applied to other fibres like jute, hemp, ramie, and raffia.

Bee Tree

The flowers of the linden are rich in nectar, attracting many bees, and where stands or plantings are abundant, the honey collected has a pleasant and distinctive flavour.

Lime

Lime meaning linden tree was a confusion, a disguised form, called dissimilation in linguistics, in which an early compound like *lind-tree* was pronounced first *line-tree* then lime-tree. There was also confusion and blending with lime, the separate and unrelated citrus fruit. Both citruses, lemon and lime, go back to Arabic words *limun* and *lima*.

Linden

Lind*en*, like asp*en* and like 'the old, oak*en* bucket' was originally an adjectival form of Old English *lind* 'lime tree.' Many Indo-European languages have this root **len* whose prime meaning is 'flexible' in reference, like the basswood-linden-*tilia* labels, to flexible fibres of the inner bark. Compare Old Norse *lind*, modern German *gelinde* 'gentle' but first 'supple, flexible, soft,' Latin *lentus* 'slow' but first 'supple, soft, lazy.' Other English words containing the root are *lithe*, and perhaps *linen* and *line* as Eric Partridge suggests from an ultimate Indo-European root **li* 'flax.' This would make **len* an extension of the flax root meaning 'flexible as threads made of flax,' then of rope or cord made of other materials, like the inner bark of the linden.

Street Name

Basswood or linden makes a dense-leafed shade tree, and is grown widely in the temperate zone along streets and in parks. For example, a famous broad avenue in old Berlin lined with grand hotels and elegant cafés was Unter den Linden 'under the lime trees,' which begins at the Brandenburg Gate and runs all the way to the River Spree. As defeat pressed in upon Nazi Berlin at the close of the Second World War, the grand avenue was bombed, and Berliners had to cut down the linden trees to burn as fuel. After the war, the street was blocked by the Berlin Wall. No longer moping in the glum confines of East Berlin, Unter den Linden today is resuming its lost bustle.

SURNAMES

Linnaeus

European scientists and humanists beginning in the Renaissance and continuing during the Enlightenment often Latinized their first and last names and affixed such grand-sounding monikers to their scientific writings. The father of taxonomy and perfecter of botanical and zoological binomials to name species published his work as Carolus Linnaeus. He was also known as Carl von Linné. But his birth name in Swedish was Karl Linn (1707-1778). In his natal Småland, a province of southern Sweden full of woods, lakes, trout streams, and now world-famous glass-works, *linn* meant linden. At the cathedral in Uppsala where

Linnaeus lies buried is this terse, apt Latin epitaph: *Princeps Botanicorum* 'the prince of botanists.'

Other Surnames

Toronto Star humorist Linwood Barclay has a first name drawn from an English surname, itself based on Linwood 'lime-tree wood.' A village in Hampshire and two in Lincolnshire bear the name.

Jenny Lind (1820-1887) was known as "the Swedish nightingale" for her agile soprano. Her surname is Old Norse, sometimes now called Old Scandinavian, for 'linden tree.'

Lindane, a now discredited insecticide—it was toxic to fish—was synthesized by and named after the Dutch chemist, Teunis van der Linden 'of the linden grove.'

More Surnames

Ukrainian: Lipa 'lime tree'
Russian: Lipin, Lipkin 'lime tree'
Finnish: Lintunen 'linden or lime tree' and Pärnänen 'lime tree' (a different species)

SPECIES

Of the thirty or so species, one, basswood, is native to Canada, growing among mixed hardwood stands in deciduous forests of the Great Lakes and St. Lawrence regions, from western New Brunswick out to the grasslands of southern Manitoba. Several other species are planted as ornamentals and shade trees. First Peoples who lived around the Great Lakes made splints of basswood bark to support injured arms and legs. They also made thread and thicker twine from the inner bark for many uses including wigwam construction.

Genus: Fagus < *fagus* Latin, beech, akin to *phegos, phagos* Greek, edible oak < *phagein* Greek, to eat, referring to the edible acorns. Compare Beechnut™ chewing gum.

BEECH

Family: Fagaceae, the beech tree family < *fagus* Latin, beech

French: *hêtre à grandes feuilles* = *Fagus grandifolia* (Latin, big-leaved)

WORD LORE

Beech < *beche* Middle English < *bece* Old English, akin to Old English *bok*, German *Buche*, Latin *fagus*, Greek *phegos*. But beech's most interesting kin word relates to the fact that sticks and slabs of its hard, tough wood were incised with the carved letters of the earliest Germanic alphabet, the runes. Such a rune tablet was in Old High German *buoh*, in modern German *Buch*, in modern English, a book!

Fagus.
The Beech.

SPECIES

Of the ten species of beech, one, *Fagus grandifolia*, is native to southern Ontario, Quebec, and our Maritimes where squirrels are particularly fond of beechnuts. Southern Ontario pioneers used springy dry beech leaves as mattress stuffing. It had more bounce to the ounce than straw. They also roasted beechnuts as a coffee substitute. *Fagus sylvatica* (Latin, of the forest) is the European beech planted in Canada as an ornamental. So is Copper Beach, *Fagus sylvatica* var. *atropurpurea* (Latin, black-purple). A dye can be extracted from Copper Beach leaves.

SURNAMES

Harriet Beecher Stowe, author of *Uncle Tom's Cabin* (1852) has a middle name from the English surname which means 'one who lives beside a beech tree.'

Horst Buchholz (German, beech grove) was a minor
German film actor who appeared in British and American films
like *Tiger Bay* (1959) and *The Magnificent Seven* (1960).

Ukrainian has surnames like Grabovsky < *hrab* 'beech
tree.'

Canadian Place Names
British explorer Frederick Beechey who served with Franklin
searching for the Northwest Passage (1818) and later with Parry
(1819-1820). Beechey Lake in the Northwest Territories is
named after him, as is Cape Beechey on Ellesmere Island.
Beechey Island in Lancaster Sound is named after his father.
In Old English, Beechey was *bece gehaeg* 'beech hedge.'

Beechwood, New Brunswick, was named after the local
hardwood stands.

Quotations
The obsolete adjective *beechen* was much used by minor
English poets who frequently repined "under beechen boughs"
or "under yonder beechen shade" often eating fruit "up piled in
beechen bowl" whereupon the reader upchucked on poetaster's
page.

ॐ

Genus: Betula > *betula* Latin, birch

Family: Betulaceae, the birch tree family

French: *bouleau* < **betullus* Latin, probable spoken variant of
betula

BIRCH

WORD LORE
Birch < *birche, birk* Middle English < *birce, beorc* Old English.
Akin to Ukrainian *bereza*, German *Birke*, Old Norse *björk*,
Sanskrit *bhurja*. A very old Indo-European tree word, its root
*bherja means that birch is the 'bright' tree, a reference to its
chalky-white bark. The words *birch* and *bright* are cognates,
words stemming from the same root cluster in Indo-European:
*bhel, **bher*, **bhrek*, all of which give words related to shining

whitely, shimmering, blazing, burning. The Ukrainian word for the month of March is *berezen* 'time when the birch trees flower.'

Birching & Fascism

In English the noun became a verb, to birch. Birching was the now-discredited practice of flogging British schoolchildren with birch twigs tied into a nasty, welt-raising bundle. Pious pedagogues even had a sad little saying to accompany their sadomasochistic act: "I must send you to Birchin Lane." The cheap pun recalled an actual London street. Even in Shakespeare's time Birchin Lane was well known for second-hand clothing stores, being lined with apparel emporia where an impecunious Elizabethan swain could pick up a bargain in a maroon velvet doublet.

The punitive use of birch is ancient. The *fasces* (Latin, literally 'bundles of stout sticks') had ceremonial importance among Roman magistrates where it was carried before them in processions as a symbol of their power over limb and life. The fasces were usually birch rods tied together with a red leather thong with an axe stuck in the midst of the rods. This recalled a time in Rome's early days when criminals were flogged and scourged with the birch rods and then beheaded with the axe. In 1919, an Italian totalitarian named Benito Mussolini brought together an anti-Communist group of thugs and street rabble called *Il Fascio di Combattenti* 'the band of fighters.' Soon other *Fasci* had sprung up. By 1921 Mussolini was in parliament and founding the *Partito Nazionale Fascista* 'National Fascist Party.' He called himself *Il Duce* 'the leader' and gave hope to other political criminals of the world, including Hitler in Germany who liked the trappings of fascism, such as borrowing the birch rod and axe symbol, although privately Hitler thought Mussolini was a vulgar if deadly clown. Took one to know one. The birch has not always been happy in its political namesakes. One thinks of the John Birch Society, another mob of anti-Communist, right-wing zealots, founded in the United States in 1958 and named after a U.S.A.F. intelligence officer, supposedly the first American killed by Chinese Communists in 1945.

SPECIES

Of about sixty world species of birch, ten are native to Canada, six of them tree-like, four shrubby. Yellow Birch, *Betula alleghaniensis* (Botanical Latin, of the Allegheny Mountains), has wood that takes stain and polish well, thus finding extensive use in woodwork of all kinds. In the prairies and in parts of British Columbia and the Northwest Territories grows Water Birch, *Betula occidentalis* (Latin, western), while our Maritimes is the homestand of Grey Birch, *Betula populifolia* (Botanical Latin, with leaves like poplar). *Bouleau blanc* and *bouleau à papier* are both Canadian French terms for *Betula papyrifera*.

White Birch Uses

White Birch, Paper Birch, or Canoe Birch, is a species with many varieties and forms. The botanical binomial is *Betula papyrifera* (Latin, paper-bearing). Because of its resistance to water and decomposition, birchbark of several species has been used as a medium to write on as far back as ancient India where *bhurja*, the Sanskrit word for birch, was noted as suitable to contain the text of sacred poems. Several of the oldest extant manuscripts are on birch-bark. Birchbark was used to make a quick pair of snow glasses by many northern peoples. During trips over snow in bright sunlight, a strip was tied around the head. Two small slits made in the bark over the eyes permitted some vision. Oil of Birch with its aroma of wintergreen is still used by tanners to make Russian leather. The oil imparts an anti-mould quality that makes Russian leather a useful cover for books worth preserving.

Canoe Birch

Betula papyrifera is widespread across much of Canada and provided a smooth, waterproof shell for one of the yarest vessels ever invented by humankind. The canoe was light, easy to

Birch (Betula alba)
with fruit

repair, lasting, and resilient, and was the first transport over the inland waters of North America. Voyageurs first traded for canoes and opened up what would become Canada through trading for furs. Native peoples of the eastern woodlands traditionally made the boats in early summer when birchbark stripped easily. After long swatches of birchbark had been sized and cut, white pine, spruce, or tamarack roots were dug up and boiled taut to make the tough thread used to stitch seams. Those seams were sealed waterproof with pine resin or spruce pitch applied with a hot stick. Canoe frames and thwarts were made from cedar soaked in water so it could be bent to the required shape.

Moose Calls
The birchbark horn was a swatch of the papery bark sewn into a cone shape. The Ojibwa and other Algonkian peoples blew through the cone making the papery layers resonate and imitating the anxious foghorn basso of a female moose in heat.

Rogan
rogan < *houragon* Canadian French < *onagan* Ojibwa, bowl, container
Birchbark rogans were used by First Peoples all across northern America. They were essentially waterproof bowls, buckets, and containers used to keep food for long periods of time, sometimes being buried or cached until needed. Birchbark strips were sewn tight with spruce-root thread and the seams sealed with spruce resin. Rogans also held maple syrup, up to five gallons in one case.

Birch Drinks
"A Glass of Birch, pardner, and the first cowpoke who snickers gets a bellyful of hot lead!" Several liquid delights were made from birch. Birchsap ginger ale and birch syrup were widely known in the pioneer west. Birch wine, an old continental cordial, is made from the thin, sugary sap of *Betula alba*, European white birch. The sap is collected in March, boiled down slightly with honey, cloves, and lemon peel, and then

fermented with yeast.
Birchwater tea, an infusion
of the leaves, was once a
specific for gout and
rheumatism.

SURNAMES

Besides the obvious southern
English ones like Birch and its
northern equivalent Birk are
surnames like Birkin, from a West
Yorkshire place name meaning 'birch
wood,' Birkenshaw 'birch shaw' where shaw is an Old
English word for copse, thicket, or small wood. French
has a Bouleau surname. German and Scandinavian
languages have surnames with the root, e.g., Birchmeyer
'birch farm,' Birckholz and Birkenstock 'birch wood,' and
Bjorkstrand 'birch beach.' A surname of Byelorus is the White
Russian Beresten 'birchbark box' where for some lost reason
the ancestor was named after such an object. Russian surnames
containing the root for birch tree include Berezin and Berezov.
One Estonian surname is Kask 'birch.' Latvian has Berzins
< *berzs* 'birch' and Kalnberzins < *kaln* Latvian, mountain
+ *berzs*.

Birch leasves with cones

Ukrainian Surnames

Ukrainian has the usual type of birch last names like Berezko
= *bereza* Ukrainian 'birch' + *ko* surname suffix. But Ukrainian
also has a group of unique and sometimes humorous surnames
formed like Shakespeare was in English, namely, an imperative
followed by a noun object. Some are actual surnames; others
are names created by humorists and dramatists. Among the
actual surnames is Lupibereza 'peel the birch' indicating either
a woodsman or an ancestor who beat others or was beaten himself (?). A few others in this subgroup of playful Ukrainian
surnames:

Peciborsc 'cook the borsch'

Tovcigrecko 'stamp the buckwheat'

Vernidub 'pull out the oak'

Nepijvoda 'don't drink water' (comic name of an alcoholic ancestor ?)

Canadian Place Names
Our maps are speckled with multiple Birch Lakes, Birch Rivers, Birch Islands. But the singingest toponym is Birchy Cove on Newfoundland's east coast. Central Saskatchewan has the town of Bjorkdale, either from a Scandinavian surname, or a pioneer descriptive in Swedish or Norwegian where it can mean 'birch valley.' A summer resort in Alberta was given the birchy name of Betula Beach in 1960. Black slaves, Loyalists to King George, fled the American Revolution and arrived in Nova Scotia in 1783. Some settled at Birchtown on the edge of Shelburne. Their village name honours Brigadier-General Samuel Birch, a British commandant who gave them food and shelter before their trek north. In northern Ontario is Wigwascence Lake, an Englishing of an Algonkian decsciptive containing *wigwas-*, the root for 'birch trees.' Compare the Ojibwa *wigwasigamig* 'birch bark lodge.'

Pre-Christian Fertility Rites
The Maypole, the resin-smeared phallic tree of European spring fertility rites was often a skinned birch. See *Quotation* below. Other ceremonial relics of tree-worship lingered well into the twentieth century, as Sir James Frazer notes in *The Golden Bough*, his monumental study of folk magic and religion. In mid-spring Russian peasants went into the woods to cut down a birch sapling, dress it in women's clothes, parade it through a village, and then toss it into a stream as a charm to bring spring rain to the fields. In some parts of rural Sweden birch twigs in leaf were carried from door to door by the boys of the village. At each cottage they asked for eggs. If they received them, they put a sprig of birch leaves over the door and sang folksongs asking for fine weather and a bountiful harvest. "Seeking the May" in rural Germany involved young men of marriage age cutting down a birch tree, skinning it, greasing it, and placing at the top: sausages, cakes, and eggs. Contests were held to see who could shimmy up the Maypole to gain the prizes. In earlier

European villages the birch Maypole, representing the renewal of fertile vigour, was a fixture in the village centre where it was set and left in place all year, to be put up afresh each spring.

Quotation

HIX' PIX ARE BIRCH STIX IN HEATHEN SEX ORGIES

If English Puritans had had tabloid newspapers, that might have been a May Day headline. Here is a description of the fetching of the Maypole during the reign of Queen Elizabeth the First from the horrified pen of a puritanical scribe named Phillip Stubbes who published it in 1583 in his *Anatomie of Abuses*. Note—but be not overly concerned about—the wild vagaries of Elizabethan spelling which had not yet been made regular by the wide use of dictionaries.

> Against May, Whitsonday, or other time, all the yung men and maides, olde men and wives, run gadding over night to the woods, groves, hills, and mountains, where they spend all the night in plesant pastimes; and in the morning they return, bringing with them birch and branches of trees, to deck their assemblies withall. And no mervaile, for there is a great Lord present amongst them, as superintendent and Lord over their pastimes and sportes, namely Sathan, prince of hel. But the chiefest jewel they bring from thence is their May-pole, which they bring home with great veneration, as thus. They have twentie or fortie yoke of oxen, every oxe having a sweet nose-gay of flouers placed on the tip of its hornes, and these oxen drawe home this May-pole (this stinkyng ydol, rather) … with two or three hundred men, women, and children following it with great devotion. And thus beeing reared up … they fall to daunce about it, like as the heathen people did at the

dedication of the Idols, whereof this is perfect pattern, or rather the thing itself. I have heard it credibly reported (and that *vive voce*) by men of great gravitie and reputation, that of fortie, threescore, or a hundred maides going to the wood overnight, there have scaresly the third part of them returned home againe undefiled.

∾

BLACK GUM

Other names: Pepperidge, Sour Gum, Tupelo

Genus: Nyssa, named after *Nysa*, a Greek water nymph, because several of the species thrive in boggy habitats.

Family: Cornaceae, the dogwood family < *cornus* Latin, the cornelian cherry tree

WORD LORE

Black Gum and Sour Gum: both common names refer to the purple, berry-like fruit that is sour and bitter, and dark purple rather than black. However, many swamp birds and mammals consume the juicy fruit with gusto. And Tupelo is a honey tree for southern bees.

Like Nyssa or Nysa, all nymphs began as female spirits of place in ancient Greece. Nysa, whose name in Greek means 'stabber, piercer,' was named after Mount Nysa in ancient Boeotia, sacred to Dionysus because it was the place where, according to Greek myth, he invented wine. The mountain was named because it seemed to pierce the sky.

Tupelo is 'tree of the swamp,' the American name for *Nyssa sylvatica*, from a North American aboriginal language of the Muskogean family called Creek, where tupelo < *ito* tree + *opilwa* swamp. The Creek peoples lived in what became Alabama, Georgia, and Florida until the U.S. government compelled them to move to reservations in Oklahoma.

SPECIES

Black Gum, *Nyssa sylvatica* (Latin, of the woods < *silva* Latin, forest), is the only species native to Canada. It grows sporadically at the extreme edge of its northern range in southwestern Ontario, around Long Point on Lake Erie, and in the

Niagara Peninsula. Its rarity in Canada makes it of no commercial use, perhaps insuring that it will endure here to delight wanderers in our autumn woods when its leaves turn a rich orange-scarlet.

Place Name
At Tupelo, Mississippi, is a National Battlefield site, where during General Sherman's bloody advance toward Atlanta northern soldiers fought a key battle with Confederate troops to safeguard their lines of communication. Elvis Presley was born in Tupelo, Miss.

∾

Other name: Cascara Buckthorn

BUCKTHORN

Genus: Rhamnus < *rhamnos* Greek, spiny shrub, any of several European buckthorns

Family: Rhamnaceae, the buckthorn family

French: *le nerprun* < *niger prunus*, Late Latin (1501 A.D.), black prune tree. A European species, *Rhamnus cathartica*, is *nerprun purgatif*.

Spanish: *cascara sagrada* 'holy bark'

WORD LORE
Buckthorn is an English word more than four hundred years old, and, unlike most such terms, we know the name of the very man who coined the word and the exact printed place of its birth. Buckthorn is a loan-translation from medieval Latin by Henry Lyte, an English horticultural writer who published a translation of *Dodoens' Niewe Herball or Historie of Plantes* in 1578 where he wrote "Valerius Cordus [calls it] Cervi spina: we may well call it in English, Bucke Thorne." Some of the shrubby species have modest thorns. *Cervi spina* is Latin 'thorn of the stag' (buck, male deer).

Cascara Medical Lore

Constipated *conquistadores* first called buckthorn *la cascara sagrada* or holy bark, because the dried bark yielded a purgative godsend to invading Europeans eating strange foods in a strange land. The Spanish conquerors and explorers were most grateful that a merciful *Dios* had provided a mild laxative in the midst of what was to them the heathen hell-hole of northern Mexico and what became southern California. Perhaps cascara's discovery gave new and piquant urgency to the phrase: *vaya con Dios*, literally 'go with God.' Cascara is a laxative used in liquid or tablet form containing anthraquinones from the bark of *Rhamnus purshiana*. These are absorbed from the small intestine and carried by circulation to act on the nerves of the large intestine, stimulating gentle intestinal contractions within six to eight hours after ingestion. In the European *materia medica*, cascara replaced the more potent and dangerous purgative found in the continental species, *Rhamnus cathartica*.

SPECIES

Rhamnus purshiana is the only tree native to Canada in this genus of about ninety species of shrubs and trees. It grows at the northern end of its range in southern British Columbia. The specific epithet honours Frederick Traugott Pursh, a German botanist (1774-1820) who made important early contributions to the study of North American flora. In 1811 he moved to what became Canada to study northern plants and died at Montreal while collecting and studying botanical specimens.

SURNAME

Byelorus has a White Russian surname, Krusina 'buckthorn.' Russian has Krusinin also with the Slavic root for buckthorn.

CEDAR

Names of various species: Eastern White Cedar, Western Red Cedar: Arborvitae

Genus: Thuja < *thuia* Greek, a kind of aromatic juniper, named by the Greeks because it was used in ritual fires and sacrifices to add a resinous odour to the proceedings. Compare the Greek verb *thuein* 'to make an offering to the gods by burning (something).'

Family: Cupressaceae, the cypress family including the junipers < *cupressus* Latin, the Italian cypress tree < *kuparissos* Greek, cypress; but the word is not of Greek origin. The root (**kup*, **kuf*) is of general Mediterranean provenance, with a reflex for example in ancient Egyptian *kufi*, a cedar incense.

French: *cèdre*, *le thuya de l'est*, eastern white cedar

WORD LORE

Cedar as Word

Cedar < *cedrus* Latin < *kedros* Greek. Both the Latin and Greek words were applied to certain junipers as well. An Indo-European cognate appears in *kadagys* Lithuanian, juniper. Citrus fruits (orange, grapefruit, lemon, lime, etc.) and citron by a circuitous route of borrowing and reborrowing are also related to cedar. Latin *citrus* is akin to *cedrus*. *Citrus* meant citron tree or cedar. The Latin was borrowed into Late Greek as *kitron*. All these *citrus* and *cedar* words are perhaps from a common Mediterranean root which also appears in *seb* and *sebt*, ancient Egyptian words for cedar.

Cedar as Name

Cedar is one of those widespread and too inclusive common names in botany and horticulture. The trees called cedar in Canada, *Thuja* species, are of the arborvitae (informal) subfamily. None of the true cedar genus *Cedrus* is native to Canada. True *Cedrus* includes the Biblical cedar-of-Lebanon and the deodar.

Arborvitae: Its Canadian Origin

Arbor vitae means 'the tree of life' in Latin. Bestowed in 1558—according to one report, by the King of France—the name commemorates the first, bleak, North American winter of 1535 for explorer Jacques Cartier and his men. Ill-prepared for the rigours of the north, the white newcomers were saved from death by scurvy when aboriginal peoples showed them how to make cedar tea from the bark and foliage of eastern white cedar. The infusion was high in vitamin C, of which scurvy is a deficiency disease. Cartier sailed back to France with cedar seedlings that were grown in Paris in 1536 amid stories about the antiscorbutic wonders of this life-giving tree.

The arborvitae story made print in the works of Carolus Clusius, the humanist name of Charles de l'Écluse (1526-1609), a famous Flemish botanist, spreading the term to other European languages. In German, for example, appeared the loan-translation, *Lebensbaum* 'tree of life.' *Tulipa clusiana*, the Lady Tulip, a charming little species tulip suitable for Canadian bulb beds, is one of many plant names that honour this botanist.

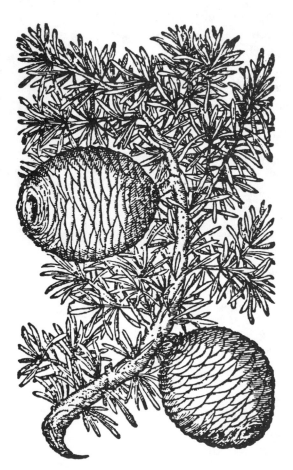

Cedar cones

SPECIES

Thuja occidentalis (Latin, western) has many confusing common names, eastern white cedar in Canada, northern white cedar in the United States, and yet note that the botanical specific means 'western'!

Thuja plicata (Latin, pleated) is western red cedar or giant Arborvitae.

Many shrub-like, horticultural varieties of Thuja, of cypress-green through golden-yellow hue, are used as hedges and ornamentals across Canada, making Thuja hybrids and dwarf cultivars among the most widely planted evergreens here. But they do not do well exposed to the pollution and dry soils of many urban sites, and often mope into scraggly decline after a season or two in a city landscape. Perhaps this is one reason plant retailers like to recommend Thuja varieties, knowing the hapless gardener will be back for more in forty-eight months?

Eastern White Cedar
Found from Manitoba to Prince Edward Island, this tree is slow growing and can live three to

four hundred years. Several living specimens on the Niagara Escarpment and elsewhere in Ontario have been proven by tree-ring dating to be twice that age. The oldest living eastern white cedar in Ontario is 1,036 years old. These are scrubby dwarf cedars growing from isolated cliffs and rockfaces where forest fires and lumber companies can't get at them. Annual growth rings on some of these cedars are one cell thick, the slowest rate of arboreal growth in the world. Shrubby forms of this tree are familiar to any outdoorsperson who has stepped gingerly to the edge of the deep, mossy bogs called cedar swamps whose spongy surfaces are firm enough to support Thuja saplings and mature but stunted specimens.

Thuja species resist drought and rotting, yet have light wood, hence their use as struts and thwarts in canoe-building. The pleasant aroma produced when Thuja boughs are burned has made white cedar important in the religious ceremonies of many First Peoples.

Aboriginals of the eastern woodlands made white cedar a plant symbol of the east. Sweetgrass symbolized the north; sage was totemic of the south; tobacco represented the west. Pioneers used cedar boughs as a strewing herb, much as we today use air-fresheners that may contain cedar-leaf oil. In the spring, musty cabins were swept with a cedar-bough broom, leaving the little home clean and fragrant.

Western Red Cedar

Thuja plicata is one of the most imposing trees of British Columbia's coastal forests and is also found inland in B.C. Well sited with moisture and rich, alluvial soil, it may soar 65 metres with stout trunks 2 metres in diameter. It is an important timber tree whose resistance to decay makes it suitable for shingles, posts, and many exterior and interior uses. One of the oldest dated speciments was eight hundred years old. Coastal peoples carved totem poles of red cedar and made dug-out canoes 20 metres long. Aeroallergens given off by western red cedar exacerbate bronchial asthma in some susceptible people. But the coastal Salish people chewed the buds of red cedar to relieve toothache. Two chemicals isolated from the wood of western red cedar are antibiotics useful against cellulose-decomposing fungi, and thus help the potency of certain wood preservatives. One is tetraphydroxystilbene. Other groups of these chemicals are called thujaplicins, after the plant's botanical name, *Thuja plicata*.

SURNAMES

Cedar-based surnames are absent from English and German, but the south of France has several last names based on trees that grew beside an ancestor's house: Cèdre, Ducèdre, and Cedras.

CHERRY

Genus: Prunus < *prunus* Latin, plum tree < *proumnos* Greek, plum tree, not a Greek word, but imported into ancient Greek, perhaps from Phrygian or some other language of Asia Minor.

Family: Rosaceae, the huge rose family

Ojibwa for black cherry: *ookwemin* 'the fruit that grows in bunches'

WORD LORE

Cherry < *chery* Middle English, from the mistaken idea that the Old Norman French form *cherise* was a plural, so one removed the *s* to get chery from cherise! Later English speakers were more accurate when they borrowed the cherry-coloured adjective from French, cerise. C*erise* French, cherry < * *ceresia* Vulgar Latin < *cerasus* Classical Latin, cherry tree < *kerasos* Greek, cherry tree. There was a town called Kerasos in Pontus in Asia Minor renowned for its cherries. The earliest form is in Akkadian, a language of ancient Mesopotamia, where the root is obvious in *karsu* 'cherry.'

Note on the Genus
Seven of the world's two hundred species are native to Canada, five cherries, two plums. See *Plum* entry. Canadians, however, grow many specimens that belong to Prunus, a large genus containing important fruit, nut, and ornamental trees, such as:

Prunus avium (Latin, of birds), the sweet cherry

Prunus amygdalus (Latin, almond), the almond

Prunus armeniaca (Latin, of Armenia), the apricot

Prunus domestica, the garden plum

Prunus persica (Latin, of Persia), the peach

Prunus triloba (Latin, with three-lobed leaves), flowering almond

Prunus yedoensis (Botanical Latin, of Yedo, older Japanese name for Tokyo), one of the many Japanese flowering cherries. *-Ensis* is a common adjectival ending in Botanical Latin; compare the Botanical Latin word for Canadian, *canadensis*.

The Cherries

Although the genus contains many important fruit trees, most parts of wild cherry trees, such as the kernel, bark, sometimes the wilted leaves, contain hydrocyanic acid and are poisonous to humans and livestock if eaten in sufficient quantity. The flesh of wild cherry fruit does not contain poison but the pits do.

PARTIAL CAUTION

Subgenus: Cerasus, sour-fruited, includes the Pin Cherry
cerasus Classical Latin, cherry tree < *kerasos* Greek, cherry tree

Subgenus: Padus, bitter-to-sweet-fruited, includes Black Cherry and Choke Cherry
padus Botanical Latin < *pados* or *pedos* Greek, the name of the European Bird Cherry

SPECIES

Black Cherry

Prunus serotina (Latin, late, in flowering and ripening) in French is *cerisier tardif* 'late cherry tree.' Black Cherry grows sparsely now across southern Ontario, Québec, parts of New Brunswick, and Nova Scotia. The only native cherry that grows to timber-sized trees, its fruit is bitter but edible. Reddish-brown Black Cherry planks once found superb use in early Canadian woodwork, especially cabinets and furniture. Black Cherry hardwood was favoured too in the early North American manufacture of butts for muskets. American names for the tree are Wild Cherry and Rum Cherry. Wild Cherry syrup extracted from the bark is still sold as a cough remedy, its use given to white settlers by first peoples who made an antitussive bark tea. Pioneers made a wine from the fermented fruit, hence Rum Cherry.

Choke Cherry
French: *cerisier de Virginie* 'cherry tree of Virginia'
Prunus virginiana (Latin, of Virginia) is usually a shrub, 3.5 metres tall, but is found occasionally as a tree, 10 metres tall. As a shrub, Choke Cherry grows throughout our Canadian provinces. Only the flesh of the bitter fruit is safe but unpalatable to eat. The common name refers to the fact that: **Choke Cherry in all its other parts is poisonous to humans and animals.**

CAUTION

Pin Cherry
Prunus pensylvanica (Botanical Latin, of Pennsylvania, and note the stubborn but accepted botanical misspelling of the state's name) can be tree-like or shrubby and is panCanadian in distribution.

Bitter Cherry
Prunus emarginata (Botanical Latin, literally 'with the border [eaten] out,' that is, the leaf edge has uneven, tooth-like projections that make it appear as though an insect has nibbled at the leaf here and there). Bitter Cherry is native to British Columbia, the northern limit of its range. First Peoples of our Pacific coast used strips of cherry and other bark to weave ceremonial baskets.

SURNAMES
Canadian hockey broadcaster Don Cherry may have an ancestor who lived beside a stand of wild cherry trees or one who grew and sold the fruit, as might persons named Cherryman and Cherriman. Also English is the surname Cherrison, which may

be a matronymic 'offspring of a woman named Cherry.' Note
that in seventeenth-century England, when Charity was a
common Christian name for women, one of the pet forms for
Charity was Cherry.

French has a surname, Cerisier, denoting an ancestor who
lived beside cherry trees. Cerise is also a last name, originally
an occupational nickname of one who sold cherries. And there
are affectionate diminutives of *cerise* in surnames like Cerison,
Cerizon, and Cerizet.

German has similar locative and occupational surnames like
Kirsch (German, cherry), Kirschbaum (cherry tree), and
Kirschenbaum.

There even appears to have been a Late Latin name *Cerisius*
that may have named a cherry-seller. The French
hamlet of Cerisy takes its name from such a *Cerisius*.

Other Surnames
Finnish: Tuomainen 'bird cherry' and Visnapu 'cherry tree'
Russian: Ceremuchin, Ceremuskin 'bird-cherry'
Ceresnev 'sweet cherry'
Visnev 'cherry'
Ukrainian: Visnja, Visnevsky < *vysnja* 'cherry tree'
Russian Jewish: Visniak 'cherry orchard'

The Cherry Tree & The Blessed Virgin Mary
Various Apocrypha recount the tale of Joseph and Mary pregnant
with Jesus wandering through an orchard. Mary asks Joseph to
pick her a cherry. Gruffly he refuses. From within the womb,
Jesus causes the cherry tree bough to bend down miraculously so
Mary can pick cherries herself. This feat of intrauterine wizardry
is perhaps wisely excluded from official Vatican lists of approved
wonderworks, but it does appear in a fifteenth-century miracle
play performed at Coventry in England and is told again in an
early English ballad "The Cherry Tree Carol."

Alcoholic Drink
Kirsch is an early Teutonic borrowing of the Latin *cerasus*,
which gives eventually a term we use in English for cherry
eau-de-vie, kirsch, a distillation from fermented cherry juice.
The German is *Kirschwasser*.

Cherry in Vulgar Slang

"She lost her cherry." Centuries old is this use of cherry as a synonym for hymen or maidenhead, although the sentence meaning 'she lost her virginity' is not recorded in print until 1928. But one hundred years earlier 'cherry pie' was an easily attainable virginal girl. From "She lost her cherry" comes a twentieth-century American college slang phrase, a cherry orchard, for a female college or girls' dormitory on campus, this use influenced by Anton Chekov's play, *The Cherry Orchard*. Elizabethan slang has *cherrylets* for a woman's nipples. Also from cherry as hymen comes its use to mean a male or female neophyte, one lacking experience: "He's a cherry. Never worked on a road crew before!" During the United States war in Vietnam, replacement troops new to the front line were called "cherries" (Hoffman, *New York Times Magazine*, December 5, 1985). In medieval England, harvest of the fruit prompted cherry fairs, notorious for hanky-panky. John Gower, a contemporary of Chaucer, wrote in *Confessio Amantis* (*A Lover's Confession*, 1393) "For all is but a chery feire." A mid-twentieth-century pop song put it this way: "Life is just a bowl of cherries." For others, "it's the pits—some other dude got all the cherries."

Proverb

>A cherry year, a merry year;
>a plum year, a dumb year;
>a pear year, a dear year.

CHESTNUT
& THE UNRELATED HORSE-CHESTNUT

Genus: Castanea < *castanea* Latin, chestnut < *kastanon* Greek, chestnut; the word is not of Greek origin but was borrowed from a language of Asia Minor; compare the Armenian *kaskeni*, chestnut tree, from *kask*, chestnut.

Family: Fagaceae, the beech family < *fagus* Latin, beech

French: *châtaigne*, the nut; *châtaignier*, the tree; and see below under *marrons glacés*.

WORD LORE

Chestnut < *chesten nut* Early Modern English < *chesten, chesteine, chastaine* Middle English < *chastaigne* Old French (*châtaigne* Early Modern French) < *castanea* Latin < *kastanea, kastanon* Greek, all meaning 'chestnut'

SPECIES

Of ten in the world, one species is native to Canada, *Castanea dentata* (Latin, toothed, referring to the leaves). Chestnut trees, once plentiful in eastern North America, were largely destroyed around the turn of the century by a fungus parasite brought in on imported Asian stock. This chestnut blight devastated the eastern North American population. Stunted Canadian specimens of *Castanea dentata* now grow sporadically only in southern Ontario.

Horse-chestnut: The Inedible Cousin, and a Different Genus
Widely grown as an ornamental in North America, the horse-chestnut is not related to *Castanea* species. *Aesculus hippocastanum* is sometimes mistaken for the native chestnut, but belongs to its own genus, Hippocastanaceae, native to northern Greece and Albania. *Aesculus* was the Latin name for an Italian oak with edible nuts, which was sacred to Jupiter. The genus and specific are Late Latin from Greek where *hippos* is 'horse.' The origin of the name is disputed. The herbalist Gerarde says "for that the people of the East countries do with the fruit thereof cure their horses of the cough." Turkish records do show the leaves and pulped nutmeat of horse-chestnuts used as a poultice for brokenwinded horses. But 'horse' prefixed to nouns in many European languages indicates something big and inferior; compare English expressions like horse-laugh, horse-mackerel, horse-mint, horse-play, and horse-radish, where there is an implied pejorative sense in the prefix.

Another suggestion, often made by Welsh writers, is that 'horse' in horse-chestnut derives from *gwres* Welsh 'fierce, pungent,' so that the horse-chestnut is the bitter one, as opposed to the sweet, edible chestnut. Still more ingenious word-sleuths have the notion that the leaf scars on horse-chestnut branches resemble perfect little horseshoes complete with seven nail markings.

Horse-chestnut Uses
British herbalists report that horse-chestnuts once found use as a folk remedy for hemorrhoids! But the practice has not spread to other lands. Horse-chestnuts are bitter-tasting and for humans inedible, as compared with sweet chestnuts, which have been a favourite European food roasted for thousands of years. However, horse-chestnuts do make handy projectiles in boys' games, hence the British children's term for them, conkers. And the older American minced oath and exclamation of incredulity "Horse chestnuts!" was a milder form of "Horse shit!"

Music Lore
Castanets, the Spanish percussion instruments, were named because they resembled a pair of chestnuts.
castanet English < *castagnette* Early Modern French
< *castañeta* Spanish < *castanetum* Latin, grove of chestnut trees

Chestnut = Old Joke
This use of chestnut, as a synonym for a tired jest or a repeated story, may have come from an otherwise justly neglected melodrama first staged at London's Covent Garden in 1816. In *The Broken Sword* by William Dimond, a tiresome old blabber named Captain Xavier is forever recounting an adventure with a tree. In the play he refers to it once as a cork-tree. His sidekick, Pablo, interrupts him to say, "A chestnut, Captain; a chestnut. I have heard you tell the joke twenty-seven times, and I am sure it was a chestnut." Perhaps, but just as cogent an origin is the knowledge among any who roast chestnuts that flavour is richest in fresh-roasted nuts. Although they are dried and stored easily, the flavour deteriorates, much like the zest in an oft-told tale.

COLOUR & FOOD LORE
Maroon as a contemporary hue describes something reddish-blue. Originally, however, the colour noun and adjective denoted a dark, shiny, reddish-brown colour, like a ripe

chestnut. It came into English from the French for Spanish chestnut, *marron*. The French first found the term in northern Italy where the Milanese dialect has *marrone* to describe a chestnut shade. *Marrone* is perhaps borrowed from a medieval Greek word, *maraon*, of obscure origin. To maroon someone on an island is not from the same source.

Marrons glacés

Large, sweet Spanish chestnuts are peeled— decorticated sounds much too surgical—gently boiled in syrup and rolled in sugar. The best *marrons glacés* (literally 'iced chestnuts') come from Privas in the Ardèche region of France.

Pane di Montagna

In some parts of rural Italy, the sweet chestnut was an important food staple, fresh-roasted or dried for storage. Because of the tree's preference for well-drained, upland soils, chestnut food was called *pane di montagna* 'mountain bread.'

PLACE NAMES & SURNAMES

Caſtanea Equina cum flore.
Horſe Cheſtnut tree in floure,

Local North American names with chestnut predate the arrival of the chestnut blight, and hark back to times when vast stands of chestnut made the wood available and prized for its beauty. In centuries before the blight, chestnuts throve in upland settings, hence frequent toponyms like Chestnut Hill and Chestnut Ridge. A long-famous trademarked brand of Canadian canoes was Chestnut. The surname is not common in English, but plentiful in French where a variety of last names occur as Châtaignier, Chatain, Chataignon, Chatignot, along with older surnames in Chastain, Chastenay, Chastagnol, etc.

Quotations

Folk Saying

"He pulled the chestnuts out of the fire." He saved the day or himself from ruin.

Poem

Perhaps the most quoted, certainly the most popular American poet of the ninetenth century was Henry Wadsworth Longfellow (1807-1882), never a versifier to miss an obvious rhyme or a memorable cadence. One of his most reprinted poems was *The Village Blacksmith*, whose first verse was:

> Under a spreading chestnut tree
> The village smithy stands;
> The smith, a mighty man is he,
> With large and sinewy hands;
> And the muscles of his brawny arms
> Are strong as iron bands.

The repetition of Longfellow's ditties called forth many parodies of his work, like this anonymous delight:

> Under a spreading gooseberry bush
> the village burglar lies,
> The burglar is a hairy man
> with whiskers round his eyes
> And the muscles of his brawny arms
> keep off the little flies.

Stalin's Chestnut Speech, January 6, 1941

The Russian mass-murderer spoke to the 8th Congress of the Communist Party and one passage of his palaver was widely reported by the world press. As usual, a liar such as Stalin was playing all ends against the middle, telling a domestic audience the opposite of what he was promising his so-called foreign allies: "The tasks of the party are … to be cautious and not allow our country to be drawn into conflicts by warmongers who are accustomed to have others pull the chestnuts out of the fire for them." While several international commentators detected the slimy deviousness of history's most lethal mustache, American Louis Fischer writing in *The Nation* magazine was succinct: "But you can burn your fingers on your own chestnuts."

Other names: Yellow Cypress, Nootka Cypress

CYPRESS

Genus: Chamaecyparis < *chamai* Greek, on the ground, hence low-growing + *cyparis* Botanical Latin, truncated form of *cyparissus* Latin, cypress < *kuparissos* Greek, cypress, not a Greek word but borrowed from some language of the Middle East. Several species of cypress are native to Persia. The Chamaecyparis genus is often called false-cypress.

Family: Cupressaceae, the cypress family

French: *cyprès*

SPECIES

The only species native to Canada, Yellow or Nootka Cypress, *Chamaecyparis nootkatensis*, hugs the British Columbia coast and its islands like the Queen Charlottes. The specific refers to Nootka Sound off Vancouver Island whose shores welcome this slow-growing tree. The Nootka are a First Nations people of the island who call themselves *Nuu Chah-nulth* 'those who live all along the mountains.' Alaska cedar, the most popular American term for this tree, illustrates how deceptive common names may be, for of course it is not a cedar.

Yellow cypress can take two hundred years to mature to a one-metre diameter and a height of 27 metres. The genus name, applied first to some prostrate, shrubby cypress species, is not too apt for large conifers. There are seven species, three native to North America, four to Japan and Taiwan.

Uses

As ornamental trees and as hedging shrubs in permanent juvenile form, false cypress species are widely planted. The long-lasting, decay-resistant wood finds use in boat building and greenhouse construction. In antiquity it was said that the gates of Byzantium (Constantinople, now Istanbul) fashioned of true cypress wood lasted 1,100 years. Now First Peoples of the Pacific coast carve cypress and use it for canoe paddles. Its very slow growth makes it appropriate for bonsai as well. Haida people used the inner bark of yellow cypress to make matting.

MYTHS

The somber-green, drooping branches of cypress trees make them seem to be musing on some immemorial melancholy. For two thousand years, Italian cypresses have stood brooding guard over cemeteries. Two twigs of cypress crossed and nailed to the door of an ancient Roman dwelling meant a recent death within. Roman funerals featured cypress boughs, and, when sufficient had been supplied to toga-clad attendees, it was, as the crafty slave Pseudolus shouts in the Broadway musical *A Funny Thing Happened on the Way to the Forum*, "Forward mourn!"

Among the numerous transformation folktales of ancient Greece, one concerned a startlingly handsome youth (were there any plain ephebes in Attica of yore?) named Kuparrisos, known to the Roman poet Ovid in his *Metamorphoses* as Cyparissus. The god Apollo fell madly in love with Cyparissus. One summer day the lad was at play in a shady dell when he killed his pet deer by mistake. Cyparissus's grief was inconsolable. He prayed to the gods to end his own life and asked Apollo to make him weep for eons. Apollo obliged his paramour by magically changing the boy into a cypress, in which form his dolefully hanging boughs would mourn forever and Cyparissus would weep resinous tears of remorse for all eternity.

Shrouded in sea fog, cypresses as sentinels of mortality beckon like lean wraiths to the newly arriving dead in Swiss painter Arnold Böcklin's *The Island of the Dead*, a now hilarious example of the funereal kitsch popular with some nineteenth-century art lovers.

Quotation

Professional gloomsters, those who have cultivated depression as others might tenderly cosset a rare orchid, and also drug-wonky renegades from the supposed tedium of normality, have, if they were writers, loved the cypress. Try Edgar Allan Poe, in the full dourness of his poem, *Ulalume*:

> Here once, through an alley Titanic,
> Of cypress, I roamed with my soul—
> Of cypress, with Psyche, my soul.

And the reader can bet that Poe's soul, which now lies fleeting, floating, yachting, boating, in the shadow of his uncoping, will be moping evermore.

∾

DOGWOOD

Other names: Western Flowering Dogwood, Pacific Dogwood

Genus: Cornus < *cornus*, Latin, name for the cornelian cherry tree. Roman armourers prized the wood of cornelian cherry

Dogwood or Cornus

because of its hardness and used it to make the hafts of spears, javelins and swords. Is this the origin of *cornus* with its similarity to *cornu* Latin, horn? Perhaps they thought the wood as hard as horn? In his classic *Native Trees of Canada*, R. C. Hosie writes that dogwood as lumber is "hard, tough, and horn-like in texture and wears smooth under friction." When early botanists came to name the cornelian tree, they too thought of it as a tough, robust member of its genus, and with habitual gender chauvinism dubbed it *Cornus mas* 'the male cornelian.' Hippocrates, the father of Greek medicine, suggested a tincture of cornelian cherries as a treatment for diarrhea. From the bark extract of an Anatolian species, *Cornus mascula*, called in Turkish *kizzilagach* 'red-wooded,' comes the red dye used to impart a traditional colour to the felt of the fez, former national head-dress of Turkish men, and still worn in Morocco.

Family: Cornaceae, the dogwood family

French: *cournouiller* < *cornualis* Latin, pertaining to horn, (later) pertaining to the cornelian tree or < **cornoialum* or some Late Latin diminutive form of Classical Latin *cornus*.

SPECIES
Two species merit special Canadian attention. One is a little wildflower called dwarf dogwood treated in the last section of this book, *Wild Plants of Canada*. Its tree-high cousin, the splendid Pacific dogwood, is the floral emblem of the province of British Columbia. *Cornus nuttallii* is for me the star tree of our southern Pacific coast and southern Vancouver Island, especially at summer's start when the larger tree forms of western flowering dogwood are spangled with starry white bracts that can measure fifteen centimetres across. The cherry and purple leaves as a background to the thousands of orange-red berries make the tree an autumn stand-out too. The specific part of the botanical name honours Thomas Nuttall (1786-1839) who was one of the pioneer botanists of the American west.

WORD LORE
Why is so fetching a family of plants saddled with that drudge of a name, dogwood? One local folk etymology of our west coast

says dogwood began life as 'dagwood,' not from the comic strip
Blondie and Dagwood, not from the 1940s Dagwood sandwich,
but from lumberman's jargon of the nineteenth century where a
dag was a wedge or skewer made from any very hard wood like
that of Pacific dogwood. That does not quite account for the fact
that we have dogwood as a word in print in England in the year
1617, referring to a related species of the British Isles, long
before dag shows up in west coast history as lumberman's slang.
Another notion from Merrie Olde England is that bark extracts
of one European species were made into a kind of shampoo
used to wash mangy dogs, hence the name. The real origin of
dogwood is more prosaic, but interesting.

For a creature often called man's best friend, the dog has
suffered much abuse, both physical and linguistic. For five
hundred years, dog, prefixed to an English word, has denoted
something worthless or contemptible. The fruit of a British
species, *Cornus sanguinea*, was called dogberry as early as
1551, possibly because the fruit was unfit for humans, but
might feed a dog. In his comedy *Much Ado About Nothing*,
Shakespeare creates the prototype of the stupid official and
dubs him Dogberry. English abounds in expressions of canine
unworthiness. To lead a dog's life was to pass through this vale
of tears as a miserable lickspittle. Dog Latin was low, illiterate,
or spurious Latin. Dog-drunk speaks for itself. Dogwood
appeared as a word after dogberry and dogtree had become
established.

Uses
Dogwood stuck as a term, even when the hardness of the wood
made it very far from useless. It was for centuries the standard
wood from which butchers' pricks were made. Prick was the
word for meat-skewer. Watchmakers cleaned out the pivot-holes
in clockwork with it. Weavers' shuttles and jewellers' blocks
were made of dogwood, and its smooth grain made it highly
suitable for toothpicks. Some piano keys are made of dogwood.
Hosie reports that "early settlers on the west coast used trunk
sections for maul and mallet heads." Animals browse the bitter
fruit of dogwood in the autumn and through the winter. Where
it will thrive, western flowering dogwood is also a popular ornamental.

Eastern Flowering Dogwood
Cornus florida has a specific that means 'flowery,' not 'from the state of Florida.' While not as showy as Pacific dogwood this is a beautiful tree in early spring bloom, with typical large white bracts around the flowers. In Canada it is confined to its northern range limit of southern Ontario. Dogwood bark found wide use among eastern First Peoples as a medicine and the source of a red dye.

SURNAMES
French has locative surnames derived from the presence of cornelian trees on an ancestor's property: Cornaille, Cornailler, Cornelier. From an eleventh-century Gallo-Roman place name *Cornualia* 'place abounding in cornelian cherry trees' comes Cornouaille. From another place name connected with this tree, Cornuejols in Aveyron, stems a similar surname.

Quotation
American humorist Irvin S. Cobb (1876-1944) mentioned the tree in a comic piece *Letter of Instructions to be opened after his death*: "Lay my ashes at the roots of a dogwood tree in Paducah at the proper planting season. Should the tree live, that will be monument enough for me." Paducah is a place name in Kentucky and in Texas.

∾

DOUGLAS FIR

See *Fir* for Balsam Fir and other Canadian species.

Genus: Pseudotsuga. *Pseudos* Greek, false + *tsu'ga* Japanese, a kind of hemlock. "False hemlock" is not an apt botanical name, for there is scant likeness between the Douglas fir and the hemlocks.

Family: Pinaceae, the pine family

French: *sapin de Douglas* < *sappinus* Late Gallic Latin, one of the European pine trees. The word seems to blend the Germanic and Scandinavian root **saf* 'sap of a plant or tree,' here the resin of the pine, **sappus* in its Gallic form, with *pinus*, Latin, pine tree. The ancient Indo-European root probably also shows reflexes in Latin words like *sapor* 'flavour' and *sapa* 'the must of wine.'

WORD LORE

Fir tree was *fyri* in Old Scandinavian, *furh* in Old English, is *fyr* in Danish and *Föhre* in modern German.

SPECIES

Douglas fir, the largest tree native to Canada, is said to reach 100 metres high. The giant coastal species of Douglas fir, *Pseudotsuga menziesii*, is one of the most valuable timber trees in the world. The common name commemorates David Douglas (1798-1834), a Scottish naturalist who was one of the first to identify this fir tree to science. As a collector of botanical specimens he travelled in Upper Canada in 1823 at Amherstburg. By 1825 he reached Fort Vancouver and then collected along the Columbia, Saskatchewan, and Hayes rivers, travelling to meet Sir John Franklin's expedition on Hudson's Bay in 1827. Two years later he returned to collect along the Okanagan and Fraser rivers. Douglas introduced the greatest number of North American plants to international botany, and some fifty species bear his name. The specific honours Archibald Menzies (1754-1842), a Scottish naval surgeon and botanical collector who sailed with Captain George Vancouver aboard *Discovery* in 1790, bringing many specimens and seeds back to England for study and classification at Kew Gardens. Menzies may have described the tree a few years before Douglas did. There is a smaller (to 140 feet) variety of Douglas fir that grows inland in British Columbia and a few places in Alberta. *Pseudotsuga menziesii* var. *glauca* is sometimes called Blue Douglas fir because its needles are grey- or bluish-green (Botanical Latin, *glaucus*).

Uses

Douglas fir is one of the best structural timbers, formerly used in bridge-building, and coated with creosote still widely used to make underwater pilings for piers, decks, and docks.

❧

ELM

Genus: Ulmus < *ulmus*, Latin, elm, from which derive the Spanish and Portuguese *olmo*. Cognate with the Latin are Scandinavian and Germanic words like Old Norse *almr*, Swedish *alm*, German *Ulme*, Dutch *olm*.

Family: Ulmaceae, the elm family

French: *orme* < *olme* Early French variant < *ulmus* Latin. Because the sounds /l/ and /r/ are phonetically similar, the change of *l* to *r* is fairly common in certain instances, and happens in many Indo-European languages.

SPECIES

Is there a Canadian town or city that does not have an Elm Street? But how many Elm Streets still have even one elm growing on them? Subdivisions of boxy bungalows with anglophilic tags like Elmwood once fringed North American cities. Few of the builders knew that elm wood was widely used to make coffins. Three species are native to Canada. The most common, White Elm, *Ulmus americana*, formerly grew in abundance from southern Saskatchewan to Cape Breton Island. Both natural and planted elms have been devastated by Dutch Elm disease, a fungus spread by European elm bark beetles introduced by accident in the early 1930s. Native insects have spread it to the whole population. Mutated and more virulent strains of the fungus have now been spread back to Europe (beginning about 1968) and have destroyed many continental specimens. Plantings of *Ulmus americana* for lawn, shade, or street trees, are no longer recommended.

Slippery Elm, *Ulmus rubra* (Latin, red, named after the rusty-red hairs on the buds) and Rock Elm, *Ulmus thomasii*, are now sparse Canadian native trees once used to make dock fenders, boat frames, and hockey sticks. Slippery elm receives its common name from the sticky

A sprig of elm leaves

inner bark that used to be processed and sold as a cough medicine. Among eastern aboriginal peoples, a powder made from Slippery Elm bark was an effective poultice for wounds, ulcers, burns, and general inflammations.

Uses

Elm rods were used by the Romans as whips to beat slaves—so frequently that Plautus, writer of Roman comedies, refers to one slave as an *ulmitriba* 'one who wears out elm whips.' When Iroquois and Ojibway peoples could not obtain birchbark, they made canoes of white elm. English boatbuilders liked water-proof elm for keel wood and bilge planks. Long before cast iron and copper were widely available, elm stems with the pith removed made early water pipes, since the wood resists decay once waterlogged. The hard wood of some European elms made them important in medieval weapon-making, which can be seen in German synonyms for *die Ulme*, terms like *Rüsternholz* 'armament wood' and *die Rüster* 'the armer.' The Wych Elm, *Ulmus glabra*, native to Scotland and northern England, is made easy to bend by steaming short boughs and once found use in bow-making. From Wych Elm diviners also chose their forked rods. The word *wych* is from *wice* Old English 'weak' and refers to the tree's pliant branches.

SURNAMES

A French soldier named Adam Dollard des Ormeaux ('of the little elm trees') is one of the martyred heroes of early New France, who died in the spring of 1660 while fighting Iroquois warriors at the siege of the Long Sault rapids. Although it was a bloody skirmish in a long and persistent war against First Peoples, Québecois historians in the nineteenth century romanticized this French aggression against native peoples, making des Ormeaux a Roman Catholic champion defending New France "from the savages." Nine of des Ormeaux's company were captured, tortured, and eaten by the victorious Iroquois. No doubt the victors remembered that "*la sauce est tout.*"

Mexican-American actor and director Edward James Olmos has a surname that began as de los Olmos 'of the elms,' denoting such trees beside an ancestral home. Delorme (French 'of the elm') is a widespread surname indicating an ancestor who dwelt beside an elm tree. The French last name Ormières was first applied to a place planted with elms.

English Elm, Elmes, Nelmes, and Thelmes are surnames denoting ancestral residence near the trees. The *n* and *th* are fragments of phrasal names that became stuck to the *elm* root; for example, *John at the Elms* was easily transformed through spoken or clerical speed into John Thelmes.

Russian surnames include Berestov and Berestin < *berest* 'elm.'

Latvian has Viksna 'elm' as a last name.

Canadian Place Names

The town of Dollard-des-Ormeaux is near Montréal. The hamlet of Elm is near Carp, Ontario, just west of Ottawa. Elm Springs bubbles in southern Saskatchewan, while Elm Creek lies southwest of Winnipeg.

Quotations

Elm. It has a sinuous, sexual, liquid sound, as American playwright Eugene O'Neill knew when he named his 1924 drama *Desire Under the Elms*, a tragedy about sexual jealousy between a father and a son.

Shakespeare refers to the Roman wine-grower's habit of training grape vines to run up elm trees that had been pollarded, that is, the vineyard planter cut off the top of young elms leaving only the trunk, from whose top sprouted shoots through which grape vines could ramble. Such pollarding of elms is still done in Portugal and Italy. In *The Comedy of Errors*, Adriana says to her mate Antipholus, "Come, I will fasten on this sleeve of thine; Thou art an elm, my husband, I a vine …"

∽

FIR

Other names: Balsam Fir, Alpine Fir, Pacific Silver Fir

Genus: Abies < *abies* Latin, fir tree

Family: Pinaceae, the pine family

French: *sapin baumier* = Balsam Fir = *Abies balsamea*

SPECIES
Species & Common Names Like Snotty Var

Balsam Fir, *Abies balsamea* (Latin, Balsam-like) grows from Newfoundland through to northern Alberta under a variety of common names. The bark on young Balsam Fir abounds with raised resin blisters, hence Blister Fir. From the imposing shape of a mature tree with its distinctive upper spire comes an early pioneer name, Church Steeple. But the most colourful slang name for the Balsam Fir was used in Newfoundland where an old fir tree with resin-clotted bark was a "snotty var," so called because it was sticky. *Var* is dialect variant of *fir*.

British Columbia's coast has related species like Pacific Silver Fir, *Abies amabilis* (Latin, lovely, referring to the attractive foliage, hence the tree's use as an ornamental). Also in British Columbia, parts of Alberta, and the Yukon is Alpine Fir, *Abies lasiocarpa* (Botanical Latin, with woolly fruit).

WORD LORE

Fir tree was *fyri* in Old Scandinavian, *furh* in Old English; is *fyr* in Danish and *Föhre* in modern German. The root abides in a common Old English collective word for woods or forest, *fyrhthe*, which produced two words encountered infrequently, unless one reads Scottish dialect poetry and woodsy English verse. In such rarified bowers of poesy a firth is a small wood, and a frith is scrub land or a hedge made of underbrush. See below under *Surnames*.

Balm of Gilead Fir

An old and incorrect name for Balsam Fir is this. Gilead was a Biblical name for the Trans-Jordan and the site of manufacture of one of the ancient Near East's most famous products. Resins from a number of local trees and shrubs—none of them *Abies balsamea*—were there made into soothing salves, skin softeners, perfumes, incense, and early cosmetics. This Balm of Gilead was expensive. In Genesis 37:25-26, Joseph's jealous brothers sell our hero to a caravan of unsavoury merchants who "came from Gilead with their camels bearing spicery and balm and myrrh, going to carry it down to Egypt."

But the most familiar citation occurs elsewhere in the Old Testament, in Jeremiah 8:22, where Holy Writ's greatest kvetch is—what else?—complaining and bitching, in this case about how evil foreign physicians are. All medicine is balderdash, implies Jeremiah, and the only healing is with God. Jeremiah employs that hoariest of devices, the rhetorical question, when he asks, "Is there no balm in Gilead?" The answer is an obvious yes. The word in the Hebrew text is *sori* and referred to a resin obtained from a small evergreen tree of the cashew family, *Pistacia lentiscus*. Yes, a sister tree, *Pistacia vera*, gives seeds sold as pistachio nuts. English translators of the Bible chose different words to render *sori*. Wyclif used "gumme." Is there no gum in Gilead? has a certain alliterative appeal. But I much prefer Coverdale's "triacle." Is there no treacle in Gilead?

A Silver Fir cone

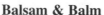

Balsam & Balm

This evocative word that now names a Canadian tree has travelled far, beginning perhaps in ancient Egypt in a word like *m'aam* 'embalming spice,' one of the spices used to mummify corpses. Note that the root still resides in our English verb *embalm*. Egyptian *m'aam* or one of its related forms like *m(b)'aam* was then borrowed by or is cognate with Hebrew *basham* and Arabic *balasan* 'balsam.' The apostrophe in *m'aam* represents a glottal stop, a common sound in Egyptian, and still in Semitic languages where it is a letter of the Hebrew and Arabic alphabets. This quick opening and closing of the glottis is often heard by persons whose languages do not possess it as a plosive, that is, as a *b* or *p* sound, thus the appearance of the *b* in dialect forms and in borrowings. Speaking of which, the Greeks borrowed the Arabic as *balsamon*, from which the Romans made *balsamum*. The Latin form evolved into Old French *basme*, giving Middle English *baume* and modern French *baume* and modern English *balm* and *balmy*. A balmy wind was originally a healthy wind. To preserve a dead body "in balm" and other spices and by additional means was, in Old French, *embasmer*, later *embaumer*, which entered English as embalm. "Not all the water in the rough, rude sea / Can wash the balm from an anointed king," crowed Shakespeare's Richard II, illustrating the Elizabethan use of balm to mean any fragrant oil or perfume.

Balsam too meant any resinous ointment that preserved, healed, or soothed. In the bizarre alchemical notions of Paracelsus, balsam was a universal preservative essence found in all organisms. In spite of his dabbling in alchemy and sundry other strands of pseudoscientific claptrap, Paracelsus did contribute to European knowledge of the medical properties of certain minerals. And we must digress to pass on the wonderfully Teutonic amplitude of his real name. He was born in 1493 and christened Theophrastus Bombastus von Hohenheim. *Ja wohl*! The word is still used in modern Italian in its curative or soothing sense: *La tua presenza è un balsamo per me* 'your being here is a comfort to me.'

Briefly the word was a verb. In J.L. Motley's *The Rise of the Dutch Republic* (1855), we read of one unfortunate who "fell down dead. We have had him balsamed and sent home." With a stern note to his relatives, no doubt, cautioning against the bad form displayed in dying abroad.

Uses

Canada Balsam is a purified resin extracted from *Abies balsamea* and still used as a glue to mount slides in microscopy. Turpentine and varnishes are made from Balsam Fir resin, by evaporation and fractional distillation. What remains afterward is a hard yellow rosin. Abietic acid (*abies* Latin, fir tree) is one of the components of resin. With its fresh balsamic aroma and needles that persist in place long after the tree is cut, Balsam Fir has been a popular Canadian Christmas tree for more than a century, with almost a million sold every year at Yuletide. Yuppie cooks now trumpet the tasty virtues of balsamic vinegar, the term being a simple translation from Italian *aceto balsamico* 'remedial vinegar,' one of the wine vinegars made by aging the must of white grapes and then flavouring it with various spices and herbs.

Sieur de Dièreville, a French surgeon and plant gatherer, visited Port Royal in 1699 to spend a year studying Acadian people and plants. In his *Relation of the Voyage to Port Royal* published in France in 1708, he reported watching aboriginal peoples resetting broken bones and then applying large pads of peat moss soaked with Balsam Fir resin (these together would have antiseptic and antibiotic properties). Birchbark was put over these pitched moss pads, then splints were tied in place with "bandages of thinner bark strips." Other early travellers in the Canadas told of bush beds made by spreading several blankets over a nest of Balsam Fir branches, "and a very soft and aromatic couch they make." Native peoples also collected Balsam Fir gum and made teas of it to treat coughs, asthma, and, less successfully, tuberculosis. The needles were sprinkled on live coals inside sweat baths and the fumes inhaled to ease cold symptoms.

Early in their food-gathering rambles, humans must have noticed Balsam Fir resin stickily coating tree wounds, buds, and cones, protecting them from infection and from voracious birds and squirrels. Fir sap is a potent natural fungicide. When the trees suffer a fungus infestation, the quantity of resin produced by an individual tree increases measurably. Fir resin protects these hardy northern conifers from cold winters too. It acts like an antifreeze for the roots, branches, and needles. Fir needles, coated with wax, can stay on the tree through their winter dormancy, and begin photosynthesis as spring thaw arrives, thus taking maximum advantage of the shorter northern growing season. The disadvantage is a forest floor strewn with resin-rich, dry needles, making vast stands of fir trees very susceptible to forest fires.

SURNAMES

English had a medieval tradesman called a balmer, one who made and sold spices and ointments. Balmer and Balme became surnames. But the English last name Balsam is deceptive, deriving from an ancestor who lived at Balsham in Cambridgeshire, whose name begins as *Baells-ham*, farm or village of an Anglo-Saxon with the common personal name *Baell*.

German has surnames like Forch, Forcher, Forchert, Forchner, and Forcht from an ancestor who lived in a *Föhrenwald*, a fir wood.

French has a wonderfully exotic matronymic (named after the mother) surname in Balmigère where the family took its name because the founding mother was an embalmer! In the Provençal language of southern France, to embalm is *balmeja*

< *basalmum* Latin, balm + *gerere* to produce, to carry. The French word for a female embalmer is *embaumeuse*.

British novelist and Edwardian esthete Ronald Firbank had a surname that meant an ancestor lived beside a wooded slope (*gefyrhthe* Old English, wooded). A whole little subclass of British last names descend from *fyrhthe*, at first meaning a group of fir trees, then any forest or woods in general. Among such surnames are Firth, Frith, Frid, Freeth, Vreede, Freak, Freke, and Firk.

Everywhere over its natural range, the fir tree has supplied surnames: Estonian Kuuzik 'fir wood,' Finnish Kuusinen 'fir tree, spruce,' Latvian Egle 'fir tree' and Eglitis 'little fir tree.'

Canadian Place Names
Toronto has its very own Balmy Beach Canoe Club. Balmy Beach near Owen Sound, Ontario has high summer breezes. As does Balm Beach, Ontario, on Nottawasaga Bay. Balsam Lake near Kirkfield, Ontario was once fringed with firs. So is Balsam Grove north of Edmonton in Alberta. The map of Canada is furry with Fir Island, Fir Mountain, Fir River, and further pinaceous toponyms.

❧

Genus: Crataegus < *krataigos* Greek, hawthorn

Family: Rosaceae, the rose family

French: *Aubépine* < *albespin* Early French, for a hawthorn called whitethorn in English < *alba spina* Latin, white thorn

HAWTHORN

SPECIES
A very large genus of short trees and coarse shrubs, hawthorns hybridize freely producing many hundreds of species. The most common Canadian hawthorn is *Crataegus chrysocarpa* (Botanical Latin, from Greek, with golden fruit) which grows from Alberta to Nova Scotia. *Crataegus douglasii* or Black Hawthorn grows as a tree in British Columbia where it was introduced into scientific record by the exploring botanist, David Douglas. Its fruit is very dark red or black. Cockspur

Hawthorn takes a small tree form in southern Québec and Ontario. Its botanical name, *Crataegus crus-galli*, includes the specific, which is Latin for 'spur of a rooster' referring to the thin, nasty thorns of this hawthorn that some early classifier thought similar to the sharp claw on the back of a rooster's lower leg. It is this spur, used as a weapon along with the beak, that makes cockfights such explosions of blood and flying feathers.

A hawthorn tree

Common Names

One still hears in Canada some of the rural English nicknames for hawthorn, such as May, Maythorn, and Mayblossom, all from its flowering time in Great Britain. European folklore opines that hawthorn, like many prickly plants, was the Crown of Thorns stuck on the head of Christ.

Hagthorn and Hedgethorn recall its ancient use as a living fence, as do Quickset and Quick. Quick here means 'alive' as opposed to fencing made of dead, cut wood. The red fruit of hawthorn has country names like Pixie Pears, Cuckoo's Beads, and Chucky Cheese. Pioneer fuel-gatherers in Canada thought the dense wood made the hottest of oven fires.

WORD LORE

Haw eaters are Canadians born and raised on Manitoulin Island. Their local word for themselves comes in three forms: run together as haweater, with a hypen as haw-eater, and primly discrete as haw eater. They like hawberries, the dark-red fruit of several species of hawthorn common in northern Ontario. Haws can be lovingly ovened in pies, tarts, and strudels. Visitors to Manitoulin buy tasty haw jams too.

The word was brought to Canada by early immigrants from England and Scotland. One of the oldest fruit words in English, haw pops up plump and ruddy in a glossary dated around A.D. 1000. Hawberry and hawthorn share an initial element that is cognate with Old High German *hag* 'enclosure.' That same root gives the modern French word for hedge, *haie*, and a living fence of hawthorn shrubs, which is *haie vive*. The first meaning of haw in English was fence. Hawthorn bushes were early used to fence yards, hence hawthorn is fence-thorn. Our later word *hedge* is related to haw, and still hemming and hawing in some rural English dialects is church-haw for churchyard.

By the fourteenth century, 'enclosed yard' and 'pen for domestic animals' were common meanings for the word *hawe*. Geoffrey Chaucer (A.D. 1340-1400), the first great poet in English, used it that way, in "The Pardoner's Tale" from his *Canterbury Tales* written in Middle English: "Ther was a polcat in his hawe, That ... hise capons hadde islawe." There was a polecat in his yard that his castrated roosters had slain [by pecking it to death]. A polecat is a smelly European weasel. Charming vignette. Chaucer used the word in its fruity sense in *The Former Age*: "They eten mast hawes and swyche pownage." They ate acorns and chestnuts (mast), hawthorn fruits, and such pannage (pig food).

A Dutch cousin of haw, Middle Dutch *hage* 'ground enclosed by a fence, park' gives both of the two names of the capital city of the Netherlands: *'s Gravenhage* 'The Count's Haw, or Park.' Modern English 'The Hague' stems directly from the other name of the city in Dutch *Den Haag* 'the hedge.' Both names refer to woods that were a royal hunting grounds surrounding a medieval palace. Some English place names whose second element is *-haw* also denote hunting parks belonging to rich landowners.

WORD LORE OF CRATAEGUS

Krataigos was the word for a species of hawthorn studied by Theophrastos, an ancient Greek botanist (371-287 B.C.), author of *The History of Plants*, one of the earliest works of botanical study based on field observation of growing things, rather than philosophical speculations.Theophrastos was the pupil and then successor of Aristotle. *Krat-* is the Greek stem meaning 'power, might, strength' which appears after being borrowed through Latin and French in many modern English words ending in *-cracy* like democracy (*democratia* Greek, rule by the people), plutocracy (*ploutokratia* Greek, rule by the rich), and aristocracy (*aristokratia* Greek, rule by the best). In his excellent *Dictionary of Plant Names for Gardeners*, W.T. Stearn, the English-speaking world's leading expert on botanical nomenclature, states that the species name Crataegus is "an allusion to the strength and hardness of the wood." As an example of scholarship about word origins, this is rather cursory and—I think—quite wrong. Yes, hawthorn wood is dense; but Stearn's etymology ignores the second root in Theophrastos's word. *Krataigos* is made up of two roots: *kratai-*'strength' + *agos* 'carrying, bringing, fetching, leading towards.' *Krataigos* means 'bringing strength' and refers to the reputed medicinal properties of various decoctions and teas made from hawthorn leaves and fruit.

Crataegus: A Second Shot at Its Etymology
A second possible etymology exists, also neglected by Stearn, in which the root of the second part of *krataigos* may be *ak* 'sharp point, thorn' with ancient Greek forms like *akis* 'point, needle, barb, hook, thorn' and *ake* 'sharp point.' The fluidity and interchange of the guttural sounds /k/ and /g/ is well-known in Indo-European languages and occurs naturally in dialects of ancient Greek, so that unattested forms like **krataiakos* shortened to **krataikos* are credible. In this case, *krataigos* then means 'strong point or thorn,' apt for almost all of the world's hundreds of hawthorn species. Compare the most common word for thorn in classical Greek, *akantha*, familiar in the acanthus leaves that decorated the capitals of Corinthian columns in Greek architecture.

Two Thorn Uses
Several species of birds called shrikes use hawthorns as killing spikes. The loggerhead shrike snatches large insects on the

wing. If the bird finds its newly acquired dainty to be a hard chew, for example, a plump beetle with a chitinous exoskeleton, the enterprising shrike finds a hawthorn bush, and impales the wriggling victim on a long, sharp thorn. Then? Yum yum! Shrikes also impale mice and smaller birds on hawthorns. Most shrikes are found in Australia, but two species show up in North America. In spite of their facial mask markings, which give them a gruesome mien, shrikes are highly beneficial birds, helping rid large parts of their range of vermin and insect pests.

Wherever hawthorns thrive, native peoples have used the larger thorns more peaceably as awls for making ornamental designs on leather.

Hawthorn as Magic Medicine

Ancient folk medicine used many plants for magical reasons, including hawthorns. The mistaken belief of primitive magic practice said, if the wood of a plant were strong, then some extract of its essence would help make an ailing human stronger. In fact, the thorn-protected haw fruits, technically pomes, are not particularly nutritious, having less than two-percent fat content, and very little sugar. Nevertheless, in many early cultures, aboriginal peoples boiled hawthorn flowers and leaves in water to make cough remedies. Boiled hawthorn roots helped back pain, again by magically "bringing strength" to the spine. Another bit of old plant magic operates here too, the doctrine of signatures, a superstitious belief summed up nicely by one eighteenth-century commentator: "There are some that think those Herbs the fittest for curing those Parts of a Man's Body, to which they bear some Sort of Resemblance, commonly called a Signature." Thus the hawthorn has thorns or spines, and must therefore participate in some mystical rapport with the human spine!

Hawthorn as Floral Symbol of Good Hope

A spring bride in ancient Greece often wore a corona of hawthorn flowers, while her *daidouchos* or torch-bearer sometimes carried a wedding torch made of hawthorn wood smeared with pine resin. Superstitious Roman mothers stuck hawthorn leaves to their newborns' cradles to ward off evil. Hawthorn flowers, beauty protected by thorns, have inspired many a poetaster to iambic bliss. Even major poets like the word, as did John Milton in *L'Allegro*, his ode to "jest and youthful jollity" where "every shepherd tells his tale / Under the hawthorn in the dale."

SURNAMES & PLACE NAMES

American writer Nathaniel Hawthorne (1804-1864) was the author of *The Scarlet Letter* (1850), *The House of the Seven Gables* (1851), and *The Marble Faun* (1860). His surname comes from one of almost a dozen English villages that arose near hawthorn thickets. German has similar surnames in Hagedorn 'haw-thorn' and Hagemann 'man who lives at a hawthorn-hedged field.'

Other English villages to produce surnames are Hawthornden and Hawthornthwaite, the latter referring to a paddock for animals enclosed by a living hawthorn fence. Four British counties have places called Hagley that gave rise to last names. One is listed in the Domesday Book for A.D.1086 as *Hageleia* 'clearing in the woods where haws (hawthorn fruits) grow.' In Suffolk is Haughley with the same meaning. The literary Brontë sisters, Anne, Charlotte, and Emily, grew up at Haworth parsonage on the West Yorkshire moors, where the place was named after "a small field enclosed with a hawthorn fence."

An occupation that produced a common English surname was that of hayward, in Old English *haege* 'hedge' + *weard* 'guard, keeper.' In a medieval village or on a manor, the hayward was a kind of bailiff who kept hedges planted and trimmed, and returned domestic livestock that had strayed. Townsfolk applied to the hayward to settle any disputes arising from use of the common. The hayward also policed the agreed-upon boundaries of grain fields given as Lammas-land. Lammas from Old English *hlafmaesse* 'loaf or bread mass' was a harvest festival held on the first of August in which loaves of bread made from the first grain harvested were blessed in the church. Lammas-land was privately-owned land made available for common use from Lammas until the next spring.

Quotation

Sir Richard Henry Bonnycastle (1791-1847) was commandant of the Royal Engineers in Upper Canada, and then in Newfoundland. Off duty, he travelled British North America and wrote charming books filled with personal observations of life in the colonies. This passage is from *Newfoundland in 1842*, published that same year in London:

> [In the environs of St. John's] the thorn or hawthorn is also found ... and would thrive extremely well if it were not for the great difficulty attendant upon their protection from cattle and goats. In winter these fences are exposed also to plunder, as no wood grows near the city, and the really heavy cost of fencing appears so lightly valued, that no law has hitherto been made to protect them, though every boy who takes his dogs to the woods to draw home fuel on a sledge ... very unceremoniously wrenches a stick or so out of the first [hawthorn] fence in his way, and will even deliberately cut down a whole panel or length, if there is a better road to be had through a field than the common highway.

I have very slightly revised this passage as quoted in *'And Some Brought Flowers': Plants in a New World* by Mary Alice Downie, Mary Hamilton, and E.J. Revell (1980).

❧

HEMLOCK

Genus: Tsuga < *tsu-ga* Japanese, hemlock. Hemlocks are worldwide in distribution through the Northern Temperate Zone. There was no clear word for hemlock in Latin or Greek, and so, happily, botanists used the name of one of the species native to northern Japan and eastern Asia. The English term *hemlock* had been given as a common name to so many different species, it was confusing and "polluted" as some botanical taxonomists testily noted. For example, there were and still are: poison hemlock (see below), water hemlock or Cicuta, and a yew called ground hemlock, *Taxus canadensis*.

Family: Pinaceae, the pine family

French: *la pruche* < Québécois French, earliest name for Eastern Hemlock, from a French dialect pronunciation of *pruce*, Early French variant of *la Prusse* 'the Prussia tree,' a spruce first identified as native to Germany. *Pruce* was also borrowed into Middle English with the *s* added later in a dialect variant to get *spruce*, all these forms ultimately < *Prussia* medieval Latin, Prussia.

SPECIES

Our Three Native Species

Tsuga canadensis, Eastern Hemlock, *la pruche de l'est, la pruche du Canada*, parts of Ontario to Prince Edward Island.

Tsuga heterophylla, Western Hemlock, chiefly British Columbia. The specific is Botanical Latin, from Greek, 'with variable leaves' referring to the fact that individual needles on a single twig can vary in length by as much as two centimetres.

Tsuga mertensiana, Mountain Hemlock, also B.C. First brought to scientific notice in 1827 by German botanical collector Karl Heinrich Mertens (1795-1830).

WORD LORE

The origin of the word *hemlock* is unknown, although its Old English noun form, *hymlice*, was first an adjective. *-Lic* is the equivalent of modern English *-like* in words such as life-like. Americans and the British tend to call the tree hemlock spruce. Canadians call it hemlock and the unrelated, lethal weed is poison hemlock.

Two facts prevented broad use of hemlock as a surname or place name: its association with poison, and the fact that at the time of the colonizing of North America, the tree was not known in England. Early colonists called it a spruce or pine. The name transfer from the European poison hemlock biennial to the North American tree happened because colonists clearing trees for farm fields burned Eastern Hemlock and thought the aroma of the needles similar to the smell of poison hemlock. Of course, a few place names squeaked by. Many North American communities have Hemlock Streets, while the United States has

Hemlock Gulch in Montana, Hemlock Mountain in Maine, and Hemlock Pass in Washington state.

Uses

The first European immigrants also noticed the smell of Tsuga needles when offered hemlock spruce tea by aboriginal peoples of the eastern woodlands, who made a brew rich in vitamin C from hemlock leaves. The Ojibway pounded into a powder the red, inner bark of *Tsuga canadensis* to heal cuts and wounds and to help stop bleeding. Needles were crushed as a spice to perk up porcupine steaks and various cuts of bear meat. A startling cinnamon-red dye extracted from the roots and inner bark of hemlock added colour to wooden spoons and other native utensils. Pioneer makers of leather extracted tannin from hemlock bark.

Hemlock gum or resin was often part of the aboriginal smoking mixture called *kinnikinick* Cree, Ojibwa 'mixture.' Kinnikinick also became a plant name for bearberry, *Arctostaphylos uva-ursi*. Smoked in a pipe, this pungent shred could be a blend of dried bearberry leaves, dried sumac leaves, red-osier dogwood bark, and tobacco.

Hemlock wood is not the strongest available but Western Hemlock is an important timber species, used for coarse lumber and for making wood pulp and plywood. Due to hemlock's dense, manifold branching, the wood has many very hard knots, hard even on axes. Tsuga species are slow growers, some to 70 metres high. The noble Methusaleh of Eastern Hemlocks was a specimen dated by dendrochronology (scientific tree-ring study) as 988 years old! It thus beat the Biblical elder by a hair, Methusaleh's mythic span being 969 years.

White-tailed deer have huddling zones called "yards" at the snow-free bases of giant forest hemlocks where they nibble the young twigs. Other munching visitors to hemlocks include moose, beaver, red squirrels, and rabbits. Also flitting in to feast are sapsuckers and siskins, chickadees and cross bills, and ruffed grouse.

Dwarf varieties of *Tsuga canadensis* are useful for hedging and garden specimens and will thrive in a shady, moist, pollution-free habitat.

Noble Forest Denizen Revealed as Hermaphrodite!

All busybody religious sects who specialize in bamboozling yokels through the promotion of sexual ignorance and intolerance, take note. Hemlock is hermaphroditic, with male and female cones on the same tree: male cones are round and tucked into the corners of the previous year's leaf axils, while oval female cones proudly jut from branch tips. Bisexual trees? Lordie, lordie, what can be next? Perhaps dedicated stitching platoons of the religious right can be organized to sew discrete hemlock skirts and save our forests from this abomination? A more perilous course would involve the acquisition of knowledge beyond Holy Writ, but that can lead to enormities like independent thought. Actually, most garden plants are hermaphroditic, in the sense that male stamens and female pistils, paired organs of reproduction, are part of each individual flower.

Socrates & The Cup of Poison Hemlock

Condemned by an Athenian jury for political reasons—officials **called** it *heresy and the corruption of youth*—the philosopher Socrates (c. 470-399 B.C.) was condemned to the customary capital punishment at that time in Athens, being forced to drink poison, as reported famously in the dialogues written by his star pupil, Plato. *The Apology* details the self-defence of Socrates at his trial; Plato's *Phaedo* recounts the death of Socrates. But Plato uses no word that means poison hemlock. It was not until more than four hundred years later that a Roman encyclopedia compiler named Pliny tells us that Socrates's fatal draught was poison hemlock extract. Plato does, however, report the poison's immediate symptoms. First Socrates' asks his jailer how precisely he should consume the poison so his death will be quick. The jailer says,"Walk around until your legs feel heavy, then lie down, and the poison will take effect." Socrates obeys. Plato reports that his beloved master lost feeling in his legs, lay down, grew cold, and died. This does not jibe with the classic symptoms of poison hemlock ingestion as noted in modern toxicological literature, which, with the victim in extremis, describes vomiting, diarrhea, gastrointestinal tract inflammation, mental confusion, wild convulsions, and violent muscular contractions caused by the operative alkaloid, coniine. Socrates's calm demeanour before death suggests not so much philosophic indifference but rather that his cup contained some potent neurotoxin and a large dose of some opiate, perhaps even Anatolian poppy extract, which Greek folk medicine knew well.

Poison Hemlock

Conium maculatum is a Eurasian biennial weed now naturalized in North America, a very poisonous member of the family Apiaceae, the celery family, quite unrelated to our conifer. *Koneion* was a Greek word for a poison plant, the *kon* root 'pine cone, pine pitch' evolved into English *cone*. But *koneion* was not necessarily the poison hemlock plant or its deadly extract. *Maculatum*, the modern specific epithet, means 'spotted' referring to stem markings, from *macula* Latin, spot, blemish, stain. Compare the immaculate conception of the Virgin Mary, immaculate because it was literally 'spotless' (*immaculatus* Ecclesiastical Latin, not spotted), that is, Mary was conceived tainted by earthly seed yes, but by a special, one-time-only dispensation of divine grace, her soul was free from original sin. This cute bit of Jesuitical sophistry keeps the womb that would bear Jesus all squeaky-clean and sinless too.

Quotation

"I have camped out, I dare say, hundreds of times, both in winter and summer, and I have never caught cold yet. I recommend, from experience, a hemlock-bed, and hemlock-tea, with a dash of whiskey in it, merely to assist the flavour, as the best preventive."

That most frequently quoted Canadian pioneer passage about hemlock is from Samuel Strickland (1804-67) who wrote of his contribution to colonial development in *Twenty-Seven Years in Canada West, or, The Experience of an Early Settler* published in 1853. Though not as famous as his two writing sisters, Susanna Moodie and Catharine Parr Traill, Strickland is worth reading.

~

HICKORY

Genus: Carya < *karya* Greek, walnut tree < *karyon* Greek, nut, kernel < **kar-* Indo-European root, hard

Family: Juglandaceae, the walnut family, includes hickories, pecans, butternuts, and walnuts, about fifty species spanning the globe in the north temperate and tropical zones. From *iuglans, iuglandis* Latin, a walnut tree sacred to Jupiter, hence the name, a contraction of *Iovis* Latin, of Jupiter + *glans, glandis* Latin, acorn, any hard nut. In ancient Greece, one of the walnut trees was sacred to Zeus and the Romans simply translated the Greek phrase *Dios balanos* 'nut of Zeus' into Latin. See *Walnut* entry.

French: *Caryer*, also *Noyer blanc d'Amérique* (American white nut tree). Continental French botany also borrowed the English word as *le hickory*.

SOME CANADIAN SPECIES

Shagbark Hickory, *Carya ovata* (Latin, oval, referring to the shape of the leaflets), is common from southern Ontario to the mouth of the St. Lawrence River. Its hickory nuts are edible. Hickories can take two hundred years to attain maturity.

Bitternut Hickory, *Carya cordiformis* (Botanical Latin, heart-shaped, referring to the nut), is the most common Canadian hickory with a range similar to Shagbark's.

Pignut Hickory, *Carya glabra* (Latin, smooth, referring to the bark on young shoots), is confined to Ontario in Canada. In the United States it is also called Smoothbark Hickory. The nuts were fed to pigs in colonial times.

Some hybridized hickories produce commercial crops of the edible nuts. There is even a hiccan, a hybrid cross between a hickory and a pecan, producing mature nuts before pecans do.

WORD LORE

Hickory came into American English first as pohickory and pokahickory, an oily paste made from hickory nuts by

Algonquin-speaking peoples of Virginia in whose language
pocohiquara or *pawcohiccora* meant 'food of pounded nuts.'
Usually made from the nuts of shagbark hickory boiled in water
and then mushed by stone, this sweet, oily, hickory "milk" was
an ingredient in colonial corn dishes like johnnycake and
hominy.

Uses

Pioneers in Ontario's Niagara peninsula made a black dye from
Mockernut Hickory, *Carya tomentosa* (Botanical Latin, hairy,
referring to the densely haired young twigs and buds).The inner
bark of Shagbark Hickory produced a yellow dye. Immigrants
also expressed nut oil for lamps from Bitternut Hickory. Pignut
Hickory wood was split into narrow bands to make pioneer
brooms, hence its other common name, Broom Hickory.

   The hickory stick terrorized pioneer schoolchildren. In the
days before corporal punishment was seen as the sadomasochistic,
inefficient, personality-damaging brutality that it is, hickory was
the wood from which the teacher's switch was made. Consider
the lyrics of this old ditty:

> School days, school days,
> Dear old golden-rule days,
> Readin' and writin' and 'rithmetic
> Taught to the tune of a hickory stick …

   Ham and bacon are still smoked over smouldering logs of
Bitternut Hickory.

   The hard wood of hickory and its ability not to splinter
under great vibration make it popular for use in the handles of
carpentry tools and skis. Early settlers made wagon shafts,
axles, and wheel spokes from the stout wood, and from hickory
bark fashioned the backs of chairs and seats.

MYTH

Dionysus, the Greek god of wine and of the irrational forces of
the mind, fell in love with a mortal maiden named Carya 'nut
tree,' daughter of a Laconian king. While Dionysus was off
presiding over frenzied rites on distant mountains, Carya died

suddenly. To honour her, Dionysus transformed the young woman into a walnut tree, at a place later called in Greek *Karuai* 'walnut trees.' The goddess Artemis appeared to the people of Laconia and explained the death of their princess, Carya. In tribute, the Laconians set up a temple dedicated to Artemis Caryatis. This, according to ancient Greek texts, was the origin of the caryatids, the female figures of stone used as supporting columns for the entablature of some Greek temples, most famously for the Erechtheum on the Athenian Acropolis, a temple named after Erechtheos, the legendary founder of Athens.

This Carya story, like the many ancient European tales of people changed into trees, hides a long-vanished rite of early Greek tree-worship. As tree-worship disappeared, the rite turned into a folk story that was garbled and altered as it was passed down by word of mouth. In origin, the Caryatids were nut-nymphs of this walnut-tree-worship, maidens who attended and performed some of the fertility rites designed to insure a bountiful harvest of nuts each year. In prehistoric European fertility ceremonies, fruit-bearing trees were, it seems, sometimes dressed in female clothing and mystically "impregnated" by a nut or fruit king selected from a tribe's mature male population. Whether or not this involved physical intromission through an aperture such as a knothole is not wholly known.

JUNIPER

Genus: Juniperus < *iuniperus* Latin, a juniper tree; the source of the compound Latin word is unknown.

Family: Cupressaceae, the cypress family

French: *le genévrier,* juniper tree; *le genièvre* juniper berry, juniper tree, and the French word for 'gin'

SPECIES

Although it has more shrub forms, botanists consider juniper the tree with the widest range in the world. Only two Canadian species attain tree size. At the northern limit of its range, confined to southern Ontario and southern Québec, is the eastern red cedar, *Juniperus virginiana* (Botanical Latin, of Virginia). The heartwood is red. It is not a true cedar but was often used to make cedar chests because of the pleasant aroma of the wood. Ornamental cultivars of this juniper are used in landscaping.

From southern Alberta's foothills to British Columbia's interior mountain ranges grows Rocky Mountain Juniper, *Juniperus scopulorum* (Latin, of the rocky cliffs). In spite of its specific, this conifer grows best in rich, well-drained, moist loam but can be seen clinging to a lofty precipice now and then.

Uses

The ripe berries—really the cones—of several juniper species are used as part of the flavouring of the French liqueur, Chartreuse. Juniper berries begin green and take two to three years to ripen to a dark blue when they are harvested to add piney zest to pâtés and to game meats like venison and rabbit. European immigrants found First Peoples burning juniper needles as incense and as an inhalant to cure coughing, as well as using the berries in herbal remedies. But note that raw, unprocessed berries of some juniper species can be toxic to livestock and are reported to have poisoned humans, although grouse, pheasants, and other birds bolt them with impunity. The cedar waxwing was actually named because of its preference for berries from the eastern red cedar.

Juniper berries

The Origin of Gin

The oldest use of juniper berries—probably six to ten thousand years old—is as a diuretic, a drug to assist the formation and excretion of urine. It was used for this purpose in ancient India and China, in primeval Europe, and by the First Peoples of the Americas. The most famous use of juniper-berry oil is to lend aroma and tang to the flavour of gin. Other herbs and spices are added to various gins, but the distinctive taste of gin was historically from an oil distilled from the berries of *Juniperus communis* (Latin, common, general, growing in groups), the shrubby juniper common to northern temperate zones around the world. The Dutch invented gin in the late 1500s as an easily distilled substitute for the more expensive juniper-berry oil. First called *genever* in Dutch (modern Dutch *jenever*), the word had been borrowed into early Dutch from Old French *geneivre*, itself from Latin *juniperus*. Returning home from wars in the Lowlands, British soldiers brought both drink and word back to England, where it sounded to British ears like the Swiss city of Geneva, hence its first appearance in English print in 1706 as Geneva. Quickly shortened to gin, the British versions of the liquor were not always flavoured with juniper berries. But, cheap to make, gin quickly caused a serious outbreak of mass alcoholism. Here from 1714 is the first use of gin in print, in *The Fable of the Bees, or Private Vices, Publick Benefits* by Bernard Mandeville: "The infamous liquor, the name of which deriv'd from Juniper-Berries in Dutch, is now, by frequent use … from a word of middling length shrunk into a Monosyllable, Intoxicating Gin." A few years later in 1738 appeared this note defining gin by poet Alexander Pope: "A spiritous liquor, the exorbitant use of which had almost destroyed the lowest rank of the People till it was restrained by an act of Parliament." But Gin Acts through the 1730s and 1740s did not curb excessive consumption. William Hogarth's moral outrage more than ten years later is proof, in the muddle of stuporous whores, gin-sodden children, and cirrhotic male imbeciles sprawled hither and thither in his famous engraving, "Gin Lane," published in 1751. Higher taxes and less potent gin helped assuage the problem toward the end of the 1750s.

Place Name
Early Acadians in Nova Scotia and New Brunswick called eastern red cedar *baton rouge* 'red stick.' After the expulsion of the Acadians and their arrival in Louisiana to become Cajuns, they found a similar tree down south and named the capital city of Louisiana after it. Western New Brunswick itself still has a little town called Juniper not too far from the Maine border. Île des Genévriers 'Juniper Tree Island' hugs the north shore of Québec a few miles from the Québec-Labrador border.

Detail from Hogarth's
"Gin Lane"

French Surnames

Several locative surnames derive from French ancestors whose habitations were near stands of juniper shrubs or near localities named after such stands. The surnames vary in spelling, depending on the part of France from which the family sprang. Examples are Genevrier, Genebrier, Genevrière, Genouvrier, Genevray, and Genevrcz.

Spanish First Name

Junípero Serra (1713-84) was a Spanish Franciscan missionary sent to open missions along the coast of what became California. The Spanish Christian name derives from Ginebro, an Italian who was a follower of Saint Francis of Assissi and whose name was latinized as *Juniperus*, because it sounded like an older Italian word for juniper, *enebro*. Compare modern

Italian *ginepro*. In fact, Ginebro is an Italian borrowing of one of the Gaelic or Welsh names like the one borne in the Camelot folktales by King Arthur's wife, Guenevere, containing the *gwyn* root that meant, when applied to humans, blond, fair-haired, or of pale complexion. The modern form comes through Cornish into English as Jennifer.

∾

MADRONA OR ARBUTUS TREE

Genus: Arbutus < Latin, the southern European strawberry tree < *arbor* Latin, tree

Family: Ericaceae, the heath family < *erice* Latin, a heath plant < *ereike* Greek, heath plant < *eri-* Greek root meaning woolly, applied to many plants with downy leaves or stems.

French: *arbousier*, the tree; *arbouse*, the fruit

Spanish: *Madroña, madroño* < *madre* Spanish, mother + *on* common augmentative suffix in the Romance languages, so that *madroña* is literally 'big mother.' It first named the European arbutus or strawberry tree whose plentiful fruits are sour but edible when sweetened.

SPECIES

Arbutus menziesii is the only Canadian species, growing on Vancouver Island and scarcely along the southern Pacific coast of British Columbia, but more plentifully down along the Oregon and California coasts. This arbutus can be identified easily in winter because it is the only broadleaved evergreen tree in Canada. Madrona also has distinctive reddish bark that peels into papery curls. The specific epithet honours Archibald Menzies, the Scottish botanist and traveller along our west coast who collected the first samples of madrona in 1792.

The European strawberry tree is *Arbutus unedo*. *Unedo* was one of the Latin names for the strawberry tree. A spurious folk etymology for *unedo* is passed on even by learned botanical taxonomists who should know better. *Unedo*, a term used by the Roman encyclopedist Pliny, is of completely unknown origin. It is certainly not a contraction of some such phrase in Latin as *unum edo* 'I eat only one' of the berries because they are so sour! This childish and lexically insupportable poppycock appears in weighty dendrology textbooks, and lessens their etymological authority each time this error is reprinted.

Uses

Arbutus menziesii grows along the Pacific coast down to California, where it is an important source of honey for California beekeepers. American common names for it include Oregon laurel and laurelwood.

Spaniards and Corsicans made a wine from the fruit of *Arbutus unedo* which was said to have a narcotic effect. So were the berries if eaten in too large a quantity. Bees love the honey-scented flowers. Roman poets knew arbutus fruit. Horace sings of the respite from Italian sun to be found in the shade of arbutus boughs; Ovid praises its plentiful "blushing fruit"; Virgil reports that goats chew arbutus shoots and country folk weave baskets from whorls of its shaggy bark.

Place Names

One of the happy isles in the coastal waters off B.C. is Madrona Island, named for this tree. Farther south in the state of Washington is the Madrona peninsula.

MAPLE

Genus: Acer < *acer, aceris* Latin, maple < **ac* Indo-European root 'sharp, pointed.' The Latin term has a cognate in the modern German word for maple, *Ahorn*, and a direct descendant in Spanish *arce*. The Latin word may refer to the hard wood. Romans used maple—among other hard woods—to make spear and pike hafts. The wooden base of a Roman child's writing tablet was often a maple board scooped out to hold the wax on which the student wrote with a stylus. But the genus name Acer may also refer to the maple's distinctively pointed leaves. Compare a synonym for the usual German *Ahorn*, *Spitzblatt* 'sharp leaf.' Russian has *ostrolistnie klejen* 'sharp-leaved maples.' Turkish borrowed the Latin *acer* to produce *akçaagac* 'maple' where *-agac* is 'tree.'

Family: Aceraceae, the maple family

French: *Érable* < *acerabulus* Vulgar Latin < *acer* Latin, maple + *abulus* possible Latin diminutive suffix, but more likely a Latin-Gallic hybrid word whose last part contains *abolo* Gallic 'rowan tree' ultimately < **abel* Indo-European root, fruit of any tree, tree-fruit. See *Apple* entry.

Quebeckers have the standard *sirop d'érable* for maple syrup, but add a near insult to define imitation syrups that are thick with glucose glop. They call this sugary impostor *sirop de poteau* 'telephone-pole syrup' or 'dead tree syrup.'

WORD LORE

Maple < *mapultreow* Old English, maple tree < **mapl-* Proto-Germanic root, maple. This appears to be a compound in which the first *m*-part's meaning is lost, but is likely the nearly world-wide **ma*, one of the first human sounds, the pursing of a baby's lips as it prepares to suck milk from mother's nipple. The **ma* root gives rise to thousands of words in many, wide-spread world languages, words like mama, mammary, mammal, *maia*, Amazon, mummy, etc. Here it would make the proto-Germanic compound **mapl-* mean 'nourishing mother tree,' that is, tree whose maple sap is nourishing. For indeed the sec-ond part of the compound, *apl-*, is a variant of Indo-European **abel* 'fruit of any tree, tree-fruit.' The primitive, etymological analogy compares the liquid sap with another nourishing liquid, mother's milk. Akin to English *maple* are Old Norse *möpurr*, Old Saxon *mapulder*, Middle Low German *mapeldorn*, and modern Irish *mailp*.

Maple Syrup Not Unique to North America

The contention of writers like Leo H. Werner in *The Canadian Encyclopedia* entry: *Maple Sugar Industry*, that maple syrup is unique to North America is suspect. China has more species of maple than any country in the world. More than one hundred species of maple are native to China, while Canada has ten native maple species. In China maple sap has been tapped for thousands of years. North America does happen to be home to the sugar maple, the species that produces the sweetest sap and the most abundant flow. But it is likely that Proto-Amerinds who crossed the Bering land bridge to populate the Americas roughly 20,000 years ago brought with them the knowledge that maple trees held a sugary sap that could be extracted for a brief period in early spring. Maple syrup may have been news to European French newcomers to North America in the seventeenth century, but it was not to Oriental peoples and to northern Europeans.

Another Indo-European Maple Word
The words for maple in the following list are cognates, literally 'born together,' but in linguistics referring to their common origin from the same Indo-European root. They all mean 'maple,' except where noted.

Ancient Greek *glinos*, *gleinos* (Theophrastos's word for the Cretan maple, *Acer creticum*)

Medieval Latin *clenus*

Russian *kljen*

Polish *klon*

Lithuanian *klevas*

Swedish *lönn*,
Norway Maple

Norwegian *lönn*,
Norway Maple

Danish *lon*,
Norway Maple
In German languages, the root *lehne, lin* was early transferred from maple to Tilia species. See *Basswood* entry.

SPECIES
Sugar Maple, *Acer saccharum* (Medieval Latin, sugar or sugary)

Soaring to thirty metres on a good site, bestowing sweetness on the spring tongue and on the autumn eye, the sugar maple grows from Lake Superior to Cape Breton Island. Acid rain threatens all maple trees and has already fatally damaged or killed thousands and thousands of specimens of this symbol of our country. Acid rain is caused chiefly by sulphur dioxide emissions from major polluters, Canadian and American industries not now policed efficiently.

Maple Emblem

Long symbolic of Canada, the leaf of a sugar maple has been the heraldic device on our flag since 1965. The Québec and Ontario coats-of-arms granted in 1868 have maple leaves; so does the 1921 Canadian coat-of-arms. But did one event begin this Canada-maple leaf association? Well, some say the maple leaf symbolism began with its use as camouflage! An intriguing suggestion, in the form of a folktale, is repeated in Frank Quance's *The Canadian Speller—Grade 6* (3rd. ed., Gage, 1950): "During the war of 1812-1814, the scarlet jacket of Canadian and British soldiers made a perfect target for the enemies. Therefore, when fighting in the woods, each soldier cut slips in his blouse and inserted a twig of maple leaves to bluff the enemy. This was the first time the maple leaf had been specifically identified with Canadians or with Canada."

One day in the fall of 1867 a Toronto school teacher named Alexander Muir was traipsing a street in the city, all squelchy underfoot from the soft felt of falling leaves, when a maple leaf alighted on his coat sleeve and stuck there. After it resisted several brushings-off, Muir joked to his walking companion that this would be "the maple leaf for *ever*!" At home that evening, he wrote a poem and set it to music, in celebration of Canada's Confederation earlier that year. Muir's song, "The Maple Leaf Forever," was wildly popular and helped fasten the symbol firmly to Canada and things Canadian.

Uses

Sugar maple is of great commercial value for its syrup and for its beautiful wood. A favourite of furniture makers, maple wood has several rare grain variations that are especially prized. Birdseye maple has a dotted pattern, while fiddleback maple and curly maple display annual rings in pleasingly undulant waves. Even maple-cured meat is a gift to white settlers' cuisine from peoples of the eastern woodlands like the Iroquois. They tell of the first human who accidentally stuck a knife in a spring maple, collected the "sweet water" and boiled some venison in this maple water.

MYTH

An Ojibwa Myth

The maple looms large in Ojibwa folktales. The time of year for sugaring-off is "in the Maple Moon." Among Ojibwa and other Algonquin-speaking peoples the mythic earth-mother and primordial female figure is Nokomis, the wise grandmother. In one story about seasonal change (among other things), cannibal wendigos, creatures of evil, chased dear old Nokomis through the countryside. Wendigos throve in icy cold. When they entered the bodies of humans, the human heart froze solid. So these wendigos represent here oncoming winter. There the evil beings were, chasing poor Nokomis, the warm embodiment of female fecundity who like the summer has grown old, pursuing her to the death. But Nokomis outsmarted the cold devils. She hid in an autumn stand of maple trees, all red and orange and deep yellow. This maple grove grew tall beside a waterfall whose mist sometimes blurred the trees' outline. And the maple trees saved Nokomis. As they peered through the mist of the waterfall, the wendigos thought they saw a raging fire of red, in which their prey was burning. But it was only old Nokomis being hidden by the bright, autumn leaves of her friends, the maples. For their service in saving the earth-mother's life, these maples were given a special gift: their water of life would be forever sweet, and humans would tap it for nourishment. All sugar maples, says the story, are descended from these trees.

Word History of *Saccharum* and *Sugar*

The specific epithet of the sugar maple is part of the history of our word *sugar*, one of the great travellers of world etymology. The Roman encyclopedist Pliny has the Classical Latin form *saccharon* for 'a sweet juice distilled from the joints of sugar cane.' The Romans borrowed the word from Greek *sakcharon*. Below, one can trace a remarkable series of word borrowings.

Verbal Timeline for Sugar
English *sugar*
Medieval French *sucre*

Italian *zucchero*
Arabic *sukkar* (Spanish *azucar* is directly from Arabic
al-sukkar with definite article assimilated)
Classical Greek *sakcharon*
Persian *shakar*
Prakrit (language of ancient India) *sakkara*
Sanskrit *sarkara*
Old Malay *singkara* (but perhaps brought there by
Buddhists whose liturgy was written in Sanskrit)

Sugarcane native to Asia came to India, then Persia, Arabia,
Africa and the near East, Phoenicia, Greece, Italy, Spain, the
rest of Europe, and finally Columbus brought it to the New
World of the Americas on his second voyage. The establishment
of sugarcane plantations led directly to the capture and enslave-
ment of black Africans. Remarkably, most of the original Asian
word clung to this sweet and dangerous gift of the earth to
humans.

Maple Medical Terms
1. **Maple Bark Disease**
Susceptible persons can get maple bark disease, an inflammation
of the lungs from inhaling spores produced by a mould,
Cryptostroma corticale, which grows in the bark of certain
maple tree species. Acute symptoms include fever, cough,
difficulty breathing, and vomiting. Chronic symptoms include
weight loss, hard breathing during exertion, and coughing up
phlegm.

2. **Maple Syrup Urine Disease**
MSUD is an inherited disorder affecting the metabolism of
certain amino acids such as valine, leucine, and isoleucine. An
enzyme needed to break these amino acids down is missing and
this metabolic flaw produces urine and sweat which smells like
maple syrup. MSUD untreated in a newborn leads to rapid
collapse of many neural functions—one symptom being exag-
gerated reflex actions—and early death. Treatment includes
peritoneal dialysis, exchange transfusion, and controlled intake
of the amino acids involved, but even these measures are not
always effective. Speed readers please pause long enough to

note that MSUD is a genetic defect. The afflicted are born with it. It is not, not, not caused by eating maple syrup!

A FEW OTHER MAPLE SPECIES

Black Maple

Acer nigrum (Latin, black) has darker bark than Sugar Maple. But Black and Sugar Maple are the woods available at the lumberyard, both sold as hard maple. Southern Ontario and Québec are the northern limits of its range.

The opening of John Gerard's treatment of maples from a 1633 Herball.

CHAP. 118. *Of the Maple tree.*

Acer maius.
The great Maple.

Acer minus.
The lesser Maple.

¶ *The Description.*

THe great Maple is a beautifull and high tree, with a barke of a meane smoothnesse : the sub-stance of the wood is tender and easie to worke on ; it sendeth forth on euery side very many goodly boughes and branches, which make an excellent shadow against the heate of the Sun

Douglas Maple or Rocky Mountain Maple

Acer glabrum var. *douglasii* is a western variety first discovered by David Douglas. See *Douglas Fir* entry. The specific refers to the smooth bark of younger trees. Douglas Maple, usually a tall shrub that turns an appealing flat red in the fall, often brightens small gardens in British Columbia where it is native to the interior and Vancouver Island. Scattered stands are found in the Alberta foothills too.

Manitoba Maple: Pollen Alert

Many a prairie street sports rows of the fast-growing Manitoba Maple, *Acer negundo*. But tree-picking town councils across Canada ought to consider the pitfall of this tree, or should I say the pollenfall. Recent scientific allergy research has shown that Manitoba Maple pollen has the highest level of allergenicity of all tree pollen **on earth**! Manitoba Maple pollen has the potential to have more susceptible people allergic to it than to any other tree. The specific *negundo* is a Sanskrit word for an Asian species called the chaste tree or *nirgundi* in Bengali and Sanskrit. An early botanic cataloguer thought the maple and the chaste tree resembled each other. Manitoba Maple grows also in Saskatchewan and Ontario, but its use as an ornamental has allowed it to spread to many areas of Canada where it is not native. An old American common name is Box Elder or Ash-leaved Maple.

Red Maple

Acer rubrum (Latin, red) is native from Newfoundland to Ontario's western border. No specific epithet could be apter: its twigs, buds, flowers, leaf-stalks, and autumn leaves are a bright tunic red.

SURNAMES

Russian Klenov is the equivalent of an English surname like Mapleson or Maples. German has two maple surnames: Ahorn and Ahorner. But the most recently notorious last name of seemingly mapley provenance belonged to the late American photographer and artist, Robert Mapplethorpe, whose obsessional snapshots of homosexual fetish objects and gargantuan penises brought a blush to many a prude's cheek and a scream of "Satan!" from many a born-again homophobe. But Mapplethorpe made religious rightwingers happy by dying of AIDS. Their vindictive glee at his death is just what Jesus ordered. Or perhaps it does not jibe with "Love thy neighbour"? It probably does not matter to the Christian Right who define themselves over and over not by what they love, but by how encompassingly they hate. In any case, Mapplethorpe as surname is not from a maple root. His surname derives from the village of Mablethorpe in Lincolnshire, listed in the Domesday Book of 1086 as *Malbertorp* and therefore from the Old Germanic personal name *Malbert* 'Gentle-Bright' + *thorp* Old Norse, outlying farm dependent on some primary settlement site, then later, village or hamlet.

Samara, the Botanical Word for Maple Key

The dry, winged seeds of maples (and a few other genus like elm) are samaras, from the Latin word *samara* which Pliny and Columella, a writer on Roman agriculture, used to mean 'elm-tree seed.' Samara is a Latin dialect version of an earlier form like **semera* related directly to the 'seed' words in Latin like *semen*. Further back in the verbal timespan, samara derives from the Indo-European root **se* and its extension **sem*, roots that mean 'one and the same.' Thus the metaphor behind sowing seeds was basically 'making more of the *same* living thing.' Children make keen whistles from samaras, best caught on the wing as the seeds helicopter down to spiral themselves into any soft earth below.

Quotation

As the author, I hereby permit myself to include one really corny quotation. This is the first poem I remember memorizing as a child, to be recited in class at Dunnville Public School. I probably received a good mark. My father, Alfred Casselman, was the principal. I can't find the poem in any anthology, and don't recall the title; but two stanzas waft back into what's left of my mind every fall when the leaves turn in Dunnville. Anyone who knows the poet and the full text is asked to drop me a note at the address given at the end of the preface.

When autumn leaves come drifting down upon a
country way,
There's nothing half so beautiful, and nothing half
so gay,
A million merry rainbow groups of maple, beech,
and oak,
All madly dancing in the wind, like little fairy folk.

And then some golden eve'tide, when each of
dancing tircs,
The country people rake them up and light the
autumn fires.
And, oh! the smell of burning leaves upon
the frosty air!
There's never a land in all the world holds incense
half so fair.

❧

OAK

Genus: Quercus < *quercus* Latin, oak. The genus is capacious,
including more than 450 species of shrub-to-tree-size species,
important now for lumber and commercial cork from the bark
of Spanish Oak, *Quercus suber*, and formerly important for
tannic acid from oak galls and the cash crop of edible acorns
in many ranges of the oak.

Family: Fagaceae, the beech family that includes oaks, beeches,
and some of the chestnuts < *fagus* Latin, beech, akin to *phegos,
phagos* Greek, edible oak < *phagein* Greek, to eat, referring to
the edible acorns

French: chêne < *chaisne* Old French < **cassanus* Early Latin-
Gallic form of lost Celtic word, oak

WORD LORE
Oak, Acorn, & Cork
I have included here the most extensive etymology in this book,
because the *oak* and *tree* words are ancient and tenacious and
have so many reflexes in modern languages that I think you will
find their history as fascinating as I did. If not, skip ahead to the
oak proverbs and myths.

Oak was the most important of trees to speakers of our ancient mother-tongue, Indo-European. Mythology and etymology attest the primacy of oak. Many words for tree first referred only to oak. Oak was *the* tree. For example, the English noun *tree* and the adjective *true* (and possibly other words like *tar*, *tray*, *trim*, *trough*, and *durable*) derive from an IE root that meant 'oak tree.' Keep in mind as you watch IE roots roll down through the millennia that consonant shifts occur in dentals, that is, Indo-European initial \d\ sounds often appear in Germanic languages as initial \t\ sounds, but in Slavic languages retain the \d\. Hence the English *tree* and the Russian *dyerevo* 'tree' evolved from the same IE root, **derwo-* 'oak.' The guttural sounds \g\ and \k\ also interchange merrily through time.

Tree Words That Begin As Oak Words in IE Languages

Basic Indo-European root forms: ** der-, *dru-, *doreu-, *derwo-* 'oak' then 'tree' or 'wood'

Sanskrit: *daru-, dru-* wood, *drumas* tree

Avestan: *dauru* piece of wood, club. Avestan is the ancient Iranian language in which Zoroaster wrote his scriptures.

Ancient Greek: *drys* oak. Compare dryads, the oak-nymphs of Greek myth who were tree-spirits. Other related Greek words are *doru* 'tree, spear' and *dendron* 'tree' from a reduplicated form of the IE oak root, **den-drevo-*.

Modern Greek: *velanidia* oak tree. This form is from *belanidi* Demotic Greek, acorn < *balanos* Classical Greek, acorn, also the Greek word for the tip of the penis or glans, hence the modern English medical term, balanitis, inflammation of the glans.

Classical Greek: *rome* a very hard kind of oak < *ronnumi* to strengthen

Latin *Quercus*, Origin of the Word *Cork*
Quercus in Latin referred usually to the Italian oak with edible acorns, an old food tree thus sacred to the chief Roman god, Jupiter. For other reasons, see the myths recounted later in this section. Mast, the meat of various nut trees, was not merely pig

fodder. Humans ate mast for thousands and thousands of years. Acorn nutmeat appears in the mythology of every western, and many eastern peoples. The *quer-* is akin to *fir* in Germanic languages, and *quercus* is the root of the word *cork*. Cork Oak was *suber* in Latin, hence the modern botanical name for Cork Oak, *Quercus suber*, a Mediterranean oak with the happy habit of making more cork tissue after some is harvested from a living tree. The Latin *suber*, Cork Oak, may be cognate with *suphar* Greek, wrinkled skin (like bark or cork) or it may hark back to the Latin and IE root *sus* 'pig' because pigs ate the acorns.

Quercus Latin, oak > *al kurk* Moorish Arabic, the Cork Oak > *alcorque* Spanish, at first a cork-soled sandal, then Cork Oak > *kork* Dutch & Low German > *cork* Middle English

Latin: *robur*, Roman name for an oak which gave the hardest wood. *Robur* Latin, hard oak < *robus* Early Latin, this oak, whence *robustus* Latin, oaken, strong as oak < *ruber* Latin, red, because of the reddish heartwood of most oaks. Now in modern botany, *Quercus robur* is the common European oak. *Robur* gives French *rouvre*, Italian *rovere*, and Spanish *roble*.

Romance Languages

Italian: *quercia* < *quercus* Latin, oak
Italian: *rovere*, Bay Oak, Durmast (*Quercus sessifolia*) < *robur*
Portuguese: *carvalho* one species of oak
Portuguese: *roble* oak < *robur*
Spanish: *roble* oak, directly < *robur* Latin, oak or < **robellus* Vulgar Latin, little oak. Compare the popular Spanish song, "*Cinco Robles*," Five Oaks.

Germanic & Scandinavian Languages

English: tree < *treo, treow* Old English < ultimately **dru-* IE root, oak
Swedish: *träd* and *trä* 'wood'
Danish: *træ*
Norwegian: *tre*
Old Norse: *tre*

English: oak < *oke* Middle English < *ac* Old English < **aig-* IE root, a kind of oak, a different species than the **dru-* oak.
All the five forms below are cognate with oak. So is the Latin word for a mountain oak, *aesculus*, and a Greek word for a certain species of oak, *aigilops*.
German: *Eiche*
Dutch: *eik*
Swedish: *ek*
Danish: *eg*
Norwegian: *eik*

Slavic Languages
This Slavic root *dub* for oak and tree may have evolved from IE **dhumbh-* 'black, dark' in reference to the dark heartwood of oak species.
Polish: *dab*, but the vowel is nasalized to give a pronunciation like *da(n)b*. The Yiddish word for oak, *demb*, was borrowed from Polish.
Russian: *dup*
Czech: *dub*

The general Slavic root for tree, *drev-*, first meant oak.
Russian: *dyerevo*
Polish: *drzewo*
Croatian: *drvo*
Church Slavonic: *drevo*
Czech: *drevo* 'wood' and *drivi* 'timber'

Albanian: *dru* 'tree, oak'

Celtic Languages
Irish: *daur*, now *dair*
Welsh: *derven*
Breton: *dero*
Cornish: *dar*, *derven*
Manx: *darragh*
Gaelic: *daragh*

Druid

Druid derives from a Celtic *oak* word. Druids were teacher-priests of an ancient Celtic religion who worshipped oak trees in Gaul, Britain, and Ireland. Their name was first borrowed into Latin and Greek from an Old Celtic or Gaulish form like *druides*. The Gaelic *druidh* seems to be a compound of **dru* + **uid* = oak-knower.

Some Oak Words in Non Indo-European Languages

Hebrew: *alon* (used as surname, e.g., Yigael Alon)

This semantic generalization of *oak* word to *tree* word occurs in language families outside IE. Consider the modern Hebrew word for tree, *ilan*, compared to the modern Hebrew word for oak, *alon*. The root consonant sounds \l\ and \n\ are the same, showing only a slight vowel gradation, common in Semitic and many other languages, used to produce extensions of meaning and various grammatical forms of the root. To give another example, vowel gradation to alter the tense happens in certain irregular English verbs: sing, sang, sung, or swim, swam, swum.

Arabic: *ballut*. Its stem meaning is acorn. The Moors brought the word to Spain where it became the Spanish word for acorn, *bellota*.

Notes on the Word *Acorn*

The Germanic and Scandinavian languages use a diminutive or a related form of their oak root to make the word for acorn. English follows suite here with Old English *aecern*. Compare Old Norse *akarn* whose prime and sensuous meaning is oak-fruit, hence acorns, hence mast from different nut trees. Here are some other acorn words:

German *Eichel* literally 'little oak thing'
Dutch *eikel*
Swedish *ekollon*, a compound of *ek* oak + *ollon* mast, acorns. But this could be folk etymology disguising what was first a diminutive form.
Danish *agern*.

SPECIES

Ten of the more than two hundred species of oak are native to
Canada. Among them, White Oak, *Quercus alba* (Latin, white,
referring to the whitish or grey-green underside of the leaves) is
the most important lumber oak. American colonists dubbed it
Stave Oak because the dense wood made tight whiskey barrels.
Southern Ontario and southern Québec are White Oak's
northern limits.

British Columbia has one native oak, the scrubby, scarce
Quercus garryana, the Garry Oak, named by early botanical
explorer David Douglas after Nicholas Garry, a Hudson's Bay
Company official who assisted him in the 1820s. The fever-
bush family, Garryaceae, is also named after this man.

From Cape Breton Island to the western shore of Lake
Superior grows Red Oak, *Quercus rubra* (Latin, red, referring to
the dark-brown, rusty red of its autumn leaves). Red Oak, in
French *le chêne rouge*, is widely used as flooring and in
furniture. Most wild birds and mammals feed on acorns in the
fall.

Proverbs

An oak is not felled at one stroke.
(In other words, be patient.)
Every oak has been an acorn.
Great oaks from little acorns grow.
Oaks may fall when reeds stand the storm.
The willow will buy a horse before the oak will pay for
a saddle. (Willows grow fast, oaks more slowly.)

Weather Rhymes

If the oak's before the ash,
Then you'll only get a splash.
If the ash precedes the oak,
then you may expect a soak.

This rhyme concerning how much rain will fall during an
English summer refers to the leafing time of the two trees.

Beware of an oak, it draws the stroke;
Avoid an ash, it counts the flash;
Creep under the thorn, it can save you from harm.

Don't seek shelter under oak or ash trees during a thunderstorm, as they are often hit by lightning.

MYTHS

The oak was sacred to early man because its broad crown was often struck by lightning and because its acorns provided sustenance. Primitive humans thought the gods hurled lightning bolts down to earth to display their displeasure at the errant ways of *homo sapiens*. Let's face it: we've been guilt-collectors since first we hopped down from a jungle tree, squished a toad, and thus brought upon ourselves—what?—the curse of a delayed rainy season because some rain god liked toads due to the fact that he looked like one. Thunder and lightning are symbols of most chief gods in world history, and the oak tree is the divine favourite of Zeus, Jupiter, Jehovah, Thor, and the old Finnish god Ukko, whose very name means 'thunder.' All these bullying deities represent in one of their several avatars the punitive, cranky old male of the tribe, a mean, vindictive elder, disgruntled that he is old, angry at his declining power, enraged that he can't have sex as often as younger males, hence the frequent scriptural denunciations of sexual pleasure and the wild ranting about holy revenge upon those who will not obey him. All this senile, graceless unwillingness to step aside for the next generation is especially prominent in dour religious texts like the Old Testament.

Worship at the Oak of Your Choice

Everywhere in their ranges, sacred oaks were sites of prayer and prophesy. At Dodona, one of the oldest magic places in ancient Greece, Zeus whispered prophetic bits of advice in the form of wind rustling through oak leaves in the sacred grove. "Yoo-hoo, Zeus! Listen, big fellah, I didn't get that last rustle. Could you zephyr that again, Your Windiness?" Historical oracle that it was, Dodona offered variety to the jaded seeker of divine signs. If one grew weary of oak-leaf divination, one of the perky oak-priestesses would waltz over and offer to receive the words of Zeus by listening to the cooing of sacred doves or by interpreting the vibrations from large brass gongs hung on the branches of the sacred oak trees and bonged by mallet or resonant in the wind, whenever a long-distance insight from Olympus was demanded. These shrine-tenders knew they had a good thing, and introduced peppy new oracular modes every so often to keep incoming customers thronging briskly at the Dodonan gates.

First Temple: An Oak Grove

The word temple derives from *templum* Latin 'a space cut off to be holy' ultimately, like the final part of a Greek word like anatomy, from IE root **tem, *ten* 'to cut or mark off.' Likewise the Teutonic words for temple all refer to what was first an enclosure around a grove of sacred trees. The word for sanctuary among the druids was *nemus*, akin to the Latin word, *nemus* 'woodland glade, grove of trees.' Holy stands of tall oaks were the first temples, and the overarching canopy of living leaves may have given later builders of cathedrals the idea for lofty vaults that spanned and enclosed a swatch of heaven.

Knock on Wood, But Do It Lightly

'To knock on wood' for good luck was first to place a hand on an oak tree. This tree-worship was a serious religion. One can often gauge the sincerity of a religion by the nastiness of retribution meted out to those who transgress its commandments. Here is Sir James Frazer, in his monumental study of magic and religion, *The Golden Bough*: "How serious that [tree] worship was in former times may be gathered from the ferocious penalty appointed by the old German laws for such as dared to peel the bark of a standing tree. The culprit's navel was to be cut out and nailed to the part of the tree which he has peeled, and he was to be driven round and round the tree till all his guts were wound about its trunk. The intention of the punishment clearly was to replace the dead bark by a living substitute taken from the culprit." For an insight into how humans have projected themselves and their needs onto nature, including trees, and for a vivid dissection of the later Nazi 'forest cult,' I recommend *Landscape and Memory* by Simon Schama (1995).

Canadian Place Names

Oakville, Ontario was named because oak staves were made into whiskey barrels and other containers there. The tree-clad snuggery of Oak Bay overlooks Haro Strait near Victoria on

Vancouver Island. It is one of British Columbia's most British enclaves, with fine mansions, tea shops, private schools, and plummy accents floating through a Pacific afternoon of scones and cricket scores.

Oak Island's Lost Pirate Treasure
In Mahone Bay off Nova Scotia's Atlantic coast, Oak Island keeps the mystery of its gold hoard, reputedly buried by Blackbeard, Morgan, or William Kidd, all pirates. The hunt for this booty began in 1795 and has never really ceased, three treasure seekers having died on Oak Island as recently as 1965. In 1971 an underwater television camera lowered into the depths showed what purported to be three chests, a metal pick, and the skeleton of a chopped-off human hand!

The Massacre of Seven Oaks
This was one of many skirmishes between rival fur companies for trading control in what is now Manitoba and areas farther west. The Hudson's Bay Company and the Northwest Company fought many times before they amalgamated in 1821. On June 19, 1816, at Seven Oaks, a few kilometres from Fort Douglas at the confluence of the Red and Assiniboine rivers, Métis working for the Northwest Company ambushed a band of men led by Robert Semple, governor of the Hudson's Bay Company's Fort Douglas. Métis killed Semple and twenty of his underlings. The HBC's Earl of Selkirk retaliated by capturing Fort William and retaking Fort Douglas. The province of Manitoba maintains Seven Oaks House as an historic site.

Oak Surnames
Sir Harry Oakes (1874-1943) was a wealthy Ontario mine owner who moved to the Bahamas in 1935 and was mysteriously murdered at his island home. The park system of Niagara Falls, Ontario, is largely a bequest of Oakes's estate to the honeymoon city.

Annie Oakley was an American sharpshooter (1860-1926).

French
André Chénier (1762-94) was a French poet guillotined during the Reign of Terror in the French Revolution. His name is part

of a group of common French surnames based on an ancestor whose house stood beside an oak tree, in French *chêne*. Only after his death were Chénier's greatest works published, and then he received the postmortem accolade of all France, being declared the greatest writer of French classical verse since Racine. An opera in the *Verismo* style based on his life is still part of the repertoire. *Andrea Chénier* with music by Umberto Giordano premiered at La Scala in Milan in 1896.

Flemish

Jan van Eyck (of the oak), one of the great Flemish masters, was born in Flanders around 1389 A.D. His masterwork is the altarpiece of the Ghent cathedral.

Fictional Surnames

Popular Canadian novelist Mazo de la Roche (1879-1961) named a fictional family after a stout species of native tree in a series of popular romances, among which was *The Whiteoaks of Jalna*.

Italian film star Vittorio de Sica played a con man asked to impersonate a dead general during World War II in *Il Generale della Rovere* directed by Roberto Rossellini in 1959. The surname *della Rovere* indicates an ancestor who dwelt in or beside a grove of Bay Oaks, *rovere* in Italian, from Latin *robur*, Roman name for an oak which gave the hardest wood, whence also *robustus* Latin, oaken, strong as oak.

SURNAMES

English

Oak is a common English last name, with variants like Oake, Oaks, and Oke. But disguised and more interesting are variants like Noak, Noakes, Nokes, and Roke. The initial consonants might be called prepositional debris stuck to Oak in some of its early phrasal surnames, for example, when *Thomas atten Oke* (1296 A.D.) 'Thomas at the oak' becomes over time Thomas Noke. All instances of the surname Roke and some instances of Rook come from Middle English *atter Oke* 'at the oak.' Compound oak surnames include Ackroyd, a Yorkshire family name denoting an ancestor who lived at the oak (Old English *ac*) clearing (Old English *rod*, Yorkshire dialect pronunciation *royd*). A variant is Oakenroyd. Acland had a founding ancestor who lived on land where oaks grew. Oaker, Oakey, Oakford, Oakhill, Oakhurst, and Oakman are okay here too.

French

Chêne and its variants are widespread throughout France in many regional variations, among which are: Chenay, Chêneau, Cheneval ('oak-valley'), and in surnames composed of prepositional phrases or with definite articles: Duchêne, Duchesne, and Lechêne.

German

Eich, Eicher, Eichler, Eichner, Eichbaum, Eichmann, and Eickemeyer are German surnames based on an ancestor living beside oaks or selling oak lumber. An interesting "house sign" surname in German is Eichhorn, a shortening of the German word for squirrel *Eichhörnchen* 'little oak (acorn) horn,' that is, little hoarder or storer of acorns. In the days before houses were numbered, many homes and shops had animal signs to identify either the house or the business. Squirrel fur was one of the hides sold by furriers in medieval Germany, and a merchant might label his house with the painting of a squirrel; then, as the age of surnames arose, he might identify himself as *Hans zum Eichhörnchen* 'Hans at the sign of the squirrel,' which became shortened in time to Hans Eichhorn. Dutch and Flemish have similar oak surnames.

Slavic

Russian has Dubov 'oak' and Dubrovin 'oak grove' and Dubinsky 'oak-tree.' Zoludev means 'acorn' from an ancestor who sold them or who had a brown complexion and earned an acorn nickname. Dubchek is a diminutive 'little oak.' Ukrainian has surnames like Derevo 'tree' and Dubogrej 'oak-warm.'

Other Languages

Latvian has surnames like Ozols 'oak' and Ozolins 'little oak.' Finnish has Tamminen from *tammi* 'oak.' Irish has Darroch, Darragh, Daraugh, all 'oak.'

Personal Name
Spanish has Robustiano, a Christian name for men, based on San Robustiano, an Italian Catholic martyr listed in the calendar of saints, from *robustus* Latin, strong as oak.

Quotation
Fairy folks are in old oaks. So runs an ancient English rhyme, referring to oakmen, wee sprites of the wood, somewhat like the nastier Teutonic trolls. Oakmen were malignant wood dwarves who wore red toadstool caps and tempted those who walked too far into the dark woods at night with food made to look like ordinary food but actually whipped up by the oakmen from various fungi. These oakmen may be the folk memory of a pre-Celtic mushroom cult, giving an hallucinogenic twist to "a trip in the woods."

PINE

Genus: Pinus < Latin, pine, in particular *pinus* was the Umbrella Pine of Italy, *Pinus pinea*, whose pine nuts are edible and whose resin fed the flames of pine torches. One way the Romans made torches was to fasten pine knots to short poles.

Family: Pinaceae, the pine family

French: *pin*

WORD LORE
The word *pine* has the Indo-European root **pi-* or **pa-* whose basic meaning was fat, lard, grease, then any thick, sticky substance like resin or gum. The IE root had extensions like **pit-*, **pin-*, and **pim-*. Pine then is indeed the gummy tree, the resinous one. In Germanic languages the IE root gives forms like fat, *fett*, and fetid (originally the bad smell of rancid fat). In ancient Greek *pion* was an adjective that meant fat and *pimele* was lard.

SPECIES
Nine of the approximtely ninety species of pines are native to Canada. Some exotic species are widely planted as ornamentals, including Scots Pine, Austrian Pine, and Mugo Pine. Unlike

most other trees, pines do well in poor, dry soils. Soft pines include Eastern White Pine, while the hard pines or pitch pines number in their ranks Jack Pine, Lodgepole Pine, Ponderosa Pine, and Red Pine.

Eastern White Pine, *le pin blanc*, is *Pinus strobus*. In Latin *strobus* was the name of a resin-yielding tree, the word borrowed from *strobos* Greek, a whirling, hence anything twisted like a pine cone; and compare the Greek diminutive *strobile* 'pine cone.' Botany uses the diminutive as a technical term where a strobile is a cone-like structure of club mosses and horsetails. Strobile also denotes any gymnosperm cone.

Ranging from Newfoundland to the border of Manitoba, Eastern White Pine is still important as softwood lumber for window frames, doors, and trim because it does not shrink. At the turn of the nineteenth century, Eastern White Pine was Canada's largest export to Britain. Poles of now largely vanished virgin White Pine made eighty-foot masts for Britain's navy. Today in Ontario less than four hundred hectares remain of the vast stands of virgin White Pine on which Canada's eastern logging industry was built. Female black bears make sanctuary nests under the protective boughs of large Eastern White Pines.

Eastern Canada's friendliest neighbour, the state of Maine, considers Eastern White Pine the state tree, hence one of Maine's nicknames, the Pine Tree State. Maine's floral emblem is the pine cone and tassel.

Southern British Columbia and Vancouver Island are home-stands of a related softwood species, Western White Pine, *Pinus monticola* (Botanical Latin, mountain-dweller) named by David Douglas in 1825.

Red Pine, *Pinus resinosa* (Botanical Latin, full of resin), *le pin rouge*, grows from eastern Manitoba to Newfoundland. A bit harder than White Pine, it takes preservatives well and thus finds use as poles, pilings, and support timbers in building. Red Pine wood harbours more resin than other pine species, making it a tough survivor of disasters that finish off other trees. Most harmful insects give Red Pine a pass. It well resists fungi, extreme cold, and forest fires, making it a favourite for mass commercial timber plantings.

Jack Pine, *Pinus banksiana, le pin gris,* named to honour Sir Joseph Banks (1743-1820), a plant collector and supporter of British botany, long associated with the Royal Botanic Gardens at Kew, England, and president of the Royal Society of London. Banks sailed with Captain Cook to the Pacific in 1768, returning to England with thousands of collected specimens. Banks visited Newfoundland, Labrador, and adjacent areas to gather plants as well, including boughs of Jack Pine, which grows from Cape Breton Island to the Yukon as one of the defining trees of our boreal forests. *Jack Pine* is the title of a famous 1916 Canadian painting by Tom Thomson, who may have seen the model tree and the site while canoeing in Algonquin Park.

Lodgepole Pine, *Pinus contorta* (Latin, twisted), has two forms. As Shore Pine it is a short, scrubby clinger to British Columbia coastal ranges. Inland, through British Columbia, parts of Alberta, and the southern Yukon, it is thin, tall, and so straight that First Peoples early used it as the support pole in teepees and wigwams, hence the common name. As timber, the inland Lodgepole finds commercially important use as siding and pulp.

Ponderosa Pine, *Pinus ponderosa* (Latin, heavy, literally 'full of weight'), is a very important timber tree in westcoast lumbering. TV trivia freaks remember it as the eponym of the Ponderosa Ranch, a thousand-acre Nevada spread owned by the Cartwright clan on *Bonanza,* a TV western series that aired from 1959 to 1973 and starred Canada's Lorne Greene.

Origin of the Word *Lumber*
White Pine lumber is a mainstay of wood yards across Canada, so this is the place for what I hope is a pleasant digression about our word *lumber.* It harks back to Lombardy, or *Lombardia,* one of the regions of Italy. A central northern region of the republic with its hub at Milan, Lombardy is bounded on the south by the Po River, and the flat, fertile plain of the Po valley. On the north, Alpine foothills gently cup in forest hands shining Italian lakes: Maggiore, Como, Varese, Lecco. Lombardy contains in fact most of *la regione dei laghi,* the Italian lake district whose

shores hold lushly gardened private villas and elegant resorts. To the east of Lombardy are Venice and Friuli; to the west Piemonte and the bustle of Turin.

The Lombards were a Germanic tribe living in northwest Germany by the first century A.D. They skirmished with the imperial legions of Rome from time to time, but were usually peaceful, successful farmers. The Roman historian Tacitus mentions them and calls them *Langobardi*. The Germanic roots of their name are *lang* 'long' and *Barte* 'axe.' The Langobardi had a characteristically ferocious Teutonic clan name, 'men of the long axe.' Ancient word studies suggest the less frightening Germanic root *Bart* 'beard.' So they might be the Long Beards? And, instead of axes, they carried sickles? Like Father Time? I don't think so. And neither does modern etymology. There is no evidence that they called themselves *Langbarden* ('long-beards'), though you will see this supposititious flapdoodle in some continental dictionaries.

Population growth and pressure from the barbarian Huns forced the Langobardi to migrate during the fourth century into most of what is now Austria. And then they crossed the Alps and invaded northern Italy. By 572 A.D. they held every major city north of the Po.

The Kingdom of Lombardy lasted for more than two hundred years, and its power extended over much of Italy, including for a time, Rome. In southern Italy they were *Langobardi* or *Longobardi*. The word survives today as a Sicilian surname, *Longobardo*. But in the north the first element of their name was modified and they were *Lombardi*.

Eventually Lombard greed for new territory forced Pope Adrian I to form an alliance with the king of the Franks. Charlemagne smashed their armies in 774 and became king of Lombardy. In the middle ages the region was a ripped quilt of quarrelsome city states. Then Spain ruled for almost two hundred years followed by a century of Austrian dominance. France had a fling from 1796 to 1814. And finally *Lombardia* became a part of Italy in 1859.

How *Lombard* Became *Lumber*

In the middle ages the Lombards were great bankers. From about 1400 A.D. immigrants from Lombardy spread through

Europe as money-changers and bankers. In London, England, they set up money-changing shops, pawnshops, and banks, and gave their name to Lombard Street in central London which is still the chief avenue of large British banking firms. Lombard became an English word for any merchant or banker. Geoffrey Chaucer, the first great poet in English, mentions this in his *Canterbury Tales* (c. 1386), specifically in "'The Shipman's Tale":

> This merchant, which that was ful war and wys.
> Creanced hath, and payd eek in Parys
> To certain Lumbardes redy in hir hond,
> The somme of gold, and hadde of hem his bond.

Chaucer's lilting Middle English may need my prosaic and literal translation: This merchant who was wary and wise, creanced (had negotiated a loan), and paid also in Paris to certain Lombards ready in their hand (at their Paris bank) the sum of gold he owed, and had of them his bond (the bond he signed on making the loan was returned to him upon his total repayment of said loan).

For several hundred years there was a common saying in British betting slang to indicate long odds, that is, a sure thing. To the question, "What are the odds, mate?" the reply was "Lombard Street to a China orange." Lombard Street was solidly there, steadfast as its banking firms. The first, sweet, eating oranges imported from China to England were popular and not too expensive at the time this saying was coined. Betting the massed wealth of Lombard Street banks against a fruit peddler's orange was to accept insanely long odds.

In Middle English *lombard* changed to lumber, because English speakers were shortening the first vowel in everyday speech and dropping the final *d*. Lombard and then lumber referred first to a banker, a moneylender, or a pawnbroker. Although the spelling of Lombard Street remained, the pronunciation of the street's name altered. By 1668, the great British diarist Samuel Pepys writes "Lumber Street" for Lombard Street. Soon the meaning of lumber grew to encompass the pawnshop itself, and lumber meant goods stored in pawn, or junk, bits and pieces stuffed into an attic or heaped up in a pawnshop or a spare room at home. Pawnbrokers from

Lombardy introduced to England the three golden balls that long symbolized a pawn shop. It is said this trio of gilt orbs was borrowed from the escutcheon of Italy's wealthiest family, and indeed a triad of aureate globosities does adorn the coat-of-arms of the Medici family.

The lumber room in a British house is still a spare room filled with unused wooden furniture and household bric-a-brac. British colonists who immigrated to America took one use of the word with them to the New World. In North America, lumber began to mean exclusively unfinished planks, cut timber, and pieces of wood in general. And soon the medieval Italian word *lombardo* was wearing full Yankee dress as a lumberjack! A jack was a man, as any man jack of the day could have told you.

What Human Body Part Was Named After a Pine Cone?
The Pineal Gland is a small, cone-shaped gland anchored to the side of the third ventricle in the centre of the brain where it secretes two hormones, melatonin and serotonin, whose functions are just beginning to be understood. Serotonin is produced by many tissues. It is a vasoconstrictor important in the control of bleeding. In the brain, serotonin may assist the transmission of nerve impulses between nerves. Melatonin seems to be an endocrine inhibitor.

The root is *pinealis* Latin, shaped like a *pinea* pine cone < *pinus* a pine tree. The Greek-speaking Roman physician Galen (131-201 A.D.) is the first to report a name for the gland. Galen wrote his medical observations in Greek and he called it *konarion,* little pine cone < *konos* Greek, pine cone. Later Italian anatomists threw out the Greek term and used a Latin one, *glandula pinealis.* The French philosopher Descartes made the pineal gland famous by declaring that it was the location of the human soul. Although "there's more to us than surgeons can remove"—to quote Alan J. Lerner's deft lyric in his musical "On A Clear Day You Can See Forever"—to this date no neurological probe has returned from the wet depths of the brain bearing at its tip a wriggling soul. The pious are free to hope. Others will make life here on earth full of hope.

Pineapple First Meant Pine Cone

The familiar golden-yellow fruit of this monocotyledonous bromeliad is a pineapple only in English. Most other languages use a form of *anana* or *ananas*, the fruit's name in Guarani, an Amerindian tongue of Bolivia and Brazil. When first imported to Elizabethan England, the anana immediately reminded the British of their pine cone. An alternate name for the pine cone in Middle English had been pineapple, which was revived to name the juicy anana. But it was a popular name as well because it prevented confusion when the banana arrived in England in 1633, almost at the same time as the anana.

Fictional Name

Guido Lorenzini, using the pen name of Collodi, wrote about a man walking through the woods who found a thick pine bough that laughed and cried. He took it back to his village and gave the magical pine piece to a woodcarver named Geppetto who made a puppet who loved to get into boyish trouble. Geppetto called his creation 'pine seed' or in Italian, Pinocchio. But there is also a suggestion in the puppet's name of a common pejorative suffix in Italian. *Un posto* is a place, but *un postaccio* is a bad place, like prison or hell. The Italian word for pine tree is also used as a male given name, Pino. So Pinocchio suggests 'naughty little pine' as well as its exact denotation of 'pine seed.' *Le Avventuri di Pinocchio* was first published in 1883.

MYTH

Pitys was the ancient Greek word for pine tree. The nymph Pitys tended pine trees for grumpy old Boreas, god of the north wind. But Pitys was lured away from her nymphly duty by horny, goaty Pan with his enticing pipes. They scampered up a mountain to the piney fastnesses of a high, cold grove and thereafter Pan did not have to pine for Pitys. But Boreas seized the pitiful Pitys and threw her against a rocky ledge where she turned into a pine tree, and forever afterward when one of her boughs is broken, Pitys weeps resin drops, tears in remembrance of Pan. This rather pedestrian metamorphosis may disguise a pre-Hellenic fertility cult, of which Pitys, the fir or pine spirit, would be the tutelary goddess, perhaps involving orgiastic spring rites in the pine woods, where an old winter king like Boreas was replaced for the fresh year by a new king of the wood like Pan, the whole shebang being solemnized by an actual coupling of the new king with a maiden representing the fir goddess or pine queen.

Quotation

> We went to see two of the *voyageurs* launch the canoe for the purpose of fire-fishing. This sport is pursued by placing over the bow a bundle of bark, pine-knots full of turpentine, or other combustible wood, and then paddling slowly over the water. One man paddles, whilst the other kneels near the fire, and watching the fish as they rise to scan the strange appearance which attracts them, he, with unerring aim, darts his fish-spear into the 'victim of curiosity.'
>
> The sight of the canoes fishing by fire light is very beautiful on a dark summer night. Large sparkles are continually falling, and floating like meteors on the placid bosom of the dark lake; while the fitful blazing of the fire, the strong reflections on the dark figures in the canoe, and the stream of pencilled light which follows its wake between the observer and the shore, heighten truly the picturesque scene.

Sir Richard Henry Bonnycastle (1791-1847) was commandant of the Royal Engineers in Upper Canada, and then in Newfoundland. Off duty, he travelled British North America and wrote personal observations of life in the colonies, like the passage above from *The Canadas in 1841*, quoted in *'And Some Brought Flowers': Plants in a New World* by Mary Alice Downie, Mary Hamilton, and E.J. Revell (1980).

POPLAR & COTTONWOODS

Genus: Populus, named by Linnaeus < *populus* Latin, poplar

Family: Salicaceae, the willow family < *salix* Latin, willow tree

French: *peuplier* < *populus* Latin, poplar

SPECIES

Six of the thirty-five or so species are native to Canada, and one or more poplars grow in every province. Trembling Aspen and Largetooth Aspen are discussed under the *Aspen* entry. The soft wood of poplars is used chiefly to make pulp.

Populus alba.
The white Poplar tree.

Balsam Poplar, *Populus balsamifera* (Botanical Latin, balsam-bearing), *peuplier baumier.* For the word lore of *balsam*, see the *Fir* entry. Tacamahac is one common name for Balsam Poplar brought north as *tacamahaca* by the Spanish who had borrowed it from *tecomahiyac*, its name in Nahuatl, a group of languages indigenous to southern Mexico and Central America. The most famous speakers of Nahuatl were the Aztecs. Among the Aztecs it was any of several gummy resins from a variety of tropical trees used as a pungent incense. Later it was applied to gum from the buds of balsam poplar used to make Aztec medicines as well as incense.

First Peoples everywhere in Balsam Poplar's range from Alaska to Newfoundland made wound salve from the fragrant buds oozing resin in the spring, and they rubbed buds inside their noses as a decongestant. Aboriginal fishermen made net-floats from Balsam Poplar bark. Stands of this tree make a quick-growing windbreak on prairie farms in western Canada. Twigs of some poplar species are used as chewing sticks to clean the teeth in parts of the world where the toothbrush is not common.

Eastern Cottonwood, *Populus deltoides* (Botanical Latin from Greek, triangle-like). The specific denotes the deltoid or triangular leaves. The common name, cottonwood, refers to the myriad of tiny seeds downy with dense, cottony hairs to facilitate dispersal by winds and air currents. An early pioneer name, Necklace Poplar, calls attention to the string of seed capsules that mature in spring on little stalks along catkins that can be twenty centimetres long and resemble a string of beads. Cottonwoods are fast growing and short-lived.

In southern Ontario and in the St. Lawrence forest region, at its northern limit, the eastern cottonwood is too scarce to make useful lumber, but in the United States where this cottonwood is widespread, the wood is used for crating, plywood, matches, and pulp. It is also planted for windbreaks and shelterbelts there. A western variety, Plains Cottonwood, is used on the Canadian Prairies for similar, protective purposes.

Black Cottonwood, *Populus trichocarpa* (Botanical Latin, from Greek, with hairy fruit). Short hairs on the fruit pods supply the reason for the specific epithet, *trichocarpa*. This largest of Canadian cottonwoods is plentiful throughout British Columbia and western Alberta where its lumber makes furniture, boxes, and plywood. From the ample boles of mature black cottonwoods, Salish people and others of B.C.'s interior hewed cottonwood canoes, dugouts used along the Fraser River and in the north up to Peace River country.

Place Names

Many local names like Poplar Bluff, which are too small to attain placement even on a map, denote stands of these trees. The hollow drum of ruffed grouse wingbeats in poplar bluffs, high wooded banks at the edge of rivers and lakes, is a sound familiar to the attentive Prairie wanderer in the spring. In the winter ruffed grouse feed on poplar catkins, buds, and twigs. In the western United States, cottonwoods line many a stream bed in dry country, so the trees were welcome signs of water to westering homesteaders, and Cottonwood as a house name was extremely common and of good omen. Many local names occur too, like Cottonwood Creek and Cottonwood Gulch. The town of Cottonwood is near the Cottonwood River northeast of Quesnel, B.C. Canada has dozens of Poplar Rivers, Points, Hills, and Creeks.

∽

Other Common names: Cinnamon Wood, Smelling Stick, Ague Tree

SASSAFRAS

Genus: Sassafras; see *Word Lore* below

Family: Lauraceae, the laurel or bay-tree family < *laurus*, Latin, bay-tree, laurel. A wreath of bay leaves was used to crown triumphant generals, hence the English use in phrases like poet laureate. From the same root comes a Late Roman given name Laurentius, and hence Lawrence, Laurence, St. Lawrence River, and the adjective in phrases like Laurentian shield.

French: *Sassafras*

WORD LORE

Sassafras begins as *saxifraga*, Latin for stone-breaker, since a little plant, unrelated to the tree species we are discussing here, likes to root in patches of gravelly soil in arctic rock clefts and is at home in limestone scree as well, all over the northern regions of the planet. Whether or not a cumulative effect of such rooting is to actually split a large stone remains to be proved. It's more likely that early botanists simply noted its preferred site. Ancient books of herbal lore say it's saxifrage because an extract of the plant was once used to break up gallstones! A Spanish word for this little saxifrage, *sasefras*, was used by the sixteenth-century Spanish botanist Monardes to name the tree of our eastern seaboard. Sassafras may have been a now lost word in an Amerindian language that sounded to Spanish ears like *sasefras*. For a description of the little saxifrage, see *Purple Mountain Saxifrage* in the next section, *Wild Plants of Canada*.

SPECIES

Sassafras albidum (Latin, whitish, referring to the underside of the leaves), the only North American species, ends its range in southern Ontario. Oil of sassafras from the roots of this tree was once used to flavour root beer. Unfortunately scientists have found that safrole, a primary component of sassafras oil, is carcinogenic, and its use is banned in the United States. It causes liver cancer in rats. These rodent hepatomata have also ended the commercial use of sassafras oil in perfumes. The inner bark used to be the source of a rich orange dye. But modern toxicology has forbidden that too. Sassafras oil found in the bark causes a contact dermatitis. Sassafras tea is now a no-no, nixed by nitpickers as well.

In spite of these scientific warnings, ground sassafras leaves are still used in Louisiana and elsewhere in the American south to make the Cajun filé powder. The leaves are added to Cajun soups and help make filé gumbo. Perhaps they should not be. One safe, completely noncarcinogenic way to enjoy sassafras is simply to view the red swash of an autumn hillside canopied with its scarlet leaves.

Other common names: Saskatoon Berry, Mountain Juneberry

SERVICEBERRY

Genus: Amelanchier < *amelancier* in Provençal and the Savoy dialect, one of the trees called medlar, but more properly one of the serviceberry species of southern Europe, Snowy Mespilus, *Amelanchier ovalis*. The fruit is subject of an Italian weather rhyme for the autumn: *Per San Martino, nespole e buon vino.* On St. Martin's Day, medlars and good wine. The medlar tree's fruits are picked in October, and then set on straw for a month. When they are half rotted, medlar fruits are eaten. They look like little apples with brown skin. Not to everyone's taste. But *de gustibus non est disputandum.* When it's a question of taste, no discussion is necessary.

Family: Rosaceae, the rose family

Modern French botanical name: *Amélanchier*

Old Canadian French: *petites poires* 'little pears'

SPECIES

Thirteen species are native to Canada, many of them difficult to identify because the species hybridize freely and there is great intraspecial variety.

Saskatoon Berry, *Amelanchier alnifolia* (Botanical Latin, alder-leaved) was one of the most important berries to First Peoples who used it in pemmican, to flavour stew and soup, and just ate the sweet, juicy berries when they ripened in late July or early August. The berries can be dried like raisins for winter use. Nowadays Saskatoon berries make fine muffins, pies, puddings, tarts, and jams. The berry gave its name to the city of Saskatoon, Saskatchewan. Its Canadian range is from the Alaska border to central Ontario to sparse occurrence in western Québec. Other common names for it include Western Serviceberry and Western Shadbush. Its showy white flowers tend to open in early spring when shad go upriver to spawn.

Mountain Juneberry, *Amelanchier bartramiana* (Botanical Latin, named after William Bartram (1739-1823) who collected

Mountain Juneberry seeds and had them shipped to London for classification and introduction to scientific botany. Juneberries are as sweet as Saskatoons but drier.

Other species include *Amelanchier laevis* (Latin, smooth, referring to the leaves), the most common Canadian species.

WORD LORE

Origin of Saskatoon

Unique among Canadian cities, the pert metropolis of Saskatoon was founded as the proposed capital of an alcohol-free country. The teetotalers in question began in Ontario in 1882, and they chartered themselves as the Temperance Colonization Society. That same year they bought one hundred thousand acres (about forty thousand hectares) of land from the Dominion Government in what is now the province of Saskatchewan. By 1883 a party of settlers was eager to flee the gin-soaked inferno of Ontario. They went west by train to Moose Jaw, then trekked overland to a place the Cree Indians had named because there were many saskatoon berry trees in the vicinity. The Cree word for these succulent purple berries is *mi-sakwato-min* 'berries from the tree of many branches.'

Quotation

"Had 'berry-pemmican' at supper. That is to say, the ordinary buffalo pemmican, with Saskootoom berries sprinkled through it at the time of making,—which acts as currant jelly does with venison, correcting the greasiness of the fat by a slightly acid sweetness. Sometimes wild cherries are used instead of the Meesasskootoom-meena."
—from *Saskatchewan and the Rocky Mountains: A Diary and Narrative of Travel … in 1859 and 1860* written by James Carnegie, Earl of Southesk, published in Edinburgh in 1875, reprinted in 1969 in Edmonton.

SPRUCE

Genus: Picea < Latin, name for a tree, Pitch Pine < *pix, picis* Latin, black sticky pitch that the Romans made by boiling down pine resin. Cognates, related words, of *pix* are the Greek word for pine tree *pitys*, the Attic Greek for pitch *pitta*, and perhaps the Old English, *pich*. All these forms flow back to the Indo-

European root **pi* or **pa* which denoted fat, grease, gum, resin, sticky material. One extension of **pi* gives *pine* tree.

Family: Pinaceae, the pine family

French: *l'épinette* little fir needle < diminutive of *épine* spine, fir needle < *spina* Latin, spine, prickle. The earliest use of *épinette* to mean spruce tree is a document of *la Nouvelle France* dated 1664. Root beer in Québec is *le bière d'épinette* (literally 'spruce beer').

WORD LORE

Spruce: From Prussian to Piney to Spiffy

Borussia was the first Baltic or Slavic name for East Prussia. In German it became *Preussen*, in medieval Latin *Prussia*; in Middle English by 1386 Chaucer writes *Pruce*, borrowing it from Norman French *Pruce*, a variant of *la Prusse*, Prussia. Very soon after *Pruce* was borrowed into Middle English, the initial and still mysterious *s* was added to get *spruce*. This additional form with initial *s* is in English documents very early, by 1378. On the other hand, the first documented use of *Pruce* in the form *Pruz* is about a century earlier, on a ship's bill of lading written in England around 1300 in anglicized Norman French: "De c. de stokfisshe venaunt del Pruz, quart" which may be translated 'about a hundred Prussian stockfish, in good condition.' So Pruce and Spruce both meant Prussia in early English. By the middle of the seventeenth century, Spruce was completely replaced as the name of the country by Prussia. By then however spruce was firmly established as the name of a fir tree. As an adjective and a verb it also still carried its meaning of dressed-up and to dress up. How did these *tree* and *dressed-up* meanings come to be?

Some of the Prussian species of fir trees were thought to be the most elegantly formed, with the straightest trunks of all evergreen conifers, the best of trees for use as the masts of British ships. The British were also partial to chests made of various Prussian fir wood. Let's eavesdrop on three last wills and testaments. From 1493: "I beqwethe to Anneys my doughter a litell spruce forcer." A forcer was a chest. From

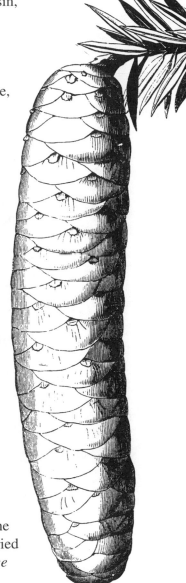

Spruce cone

1522: "I bequeathe to my said Wyffe … a spruse coffer." For over three hundred years Brits were mad-keen for any garment made of Prussian leather, the material being particularly desired for vests and overwear. Another will from 1597 leaves "my best gowne and a spruce jerkyn." In *A Briefe Description of the Whole Worlde* (1599), George Abbot wrote "the English do …bring from thence a kinde of leather, which was wont to be used in Jerkins, and called by the name of Spruce-Leather-Jerkins." A writer in 1606 envies "the sprewsest Citie-lads" for their foppish finery. In 1609 dramatist Thomas Dekker writes of "the neatest and sprucest leather." It is not far from such usage to persons sprucing up before going out for the evening, or sprucing up the cottage for the weekend.

SPECIES

Five of the forty or so spruce species are native to Canada. More lumber and pulp are made from spruce than from any other tree.

White Spruce, *Picea glauca* (Latin, bluish-green, referring to the bluish tint of the needles), *épinette blanche*, is the most widely distributed tree in North America. Some fossilized spruce is 65 million years old. Springy branches help spruce survive the burden of snow weight in deep winter. First Peoples used the flexible roots of White Spruce to sew up birchbark canoes, rogans, and other vessels required to be watertight.

Black Spruce, *Picea mariana* (Botanical Latin, of Maryland), *épinette noire*, also known as Bog Spruce or Swamp Spruce. The specific epithet was added by Philip Miller (1691-1771), writer of a useful *Dictionary of Gardens* (1731). Maryland symbolized the North American bounty of species to this British curator of the Physic Garden at Chelsea. Black Spruce does not grow in Maryland.

Sitka Spruce, *Picea sitchensis* (Botanical Latin, of Sitka, Alaska), grows all along our Pacific coast, on Vancouver Island, and thrives on the Queen Charlotte Islands. The tallest spruce, it attains heights of 60 metres, making for long, clear boards of sawn lumber, and contributing to this wood's importance as a timber-producer in B.C. The Salish, the Nootka, and other First Peoples of British Columbia make Sitka spruce baskets and craftwork.

Uses of Spruce Tea
"The Great Antiscorbutic Elixir of the Canadas, Most Easily Made and Necessary To Be Had by All Persons There Resident"

Scurvy was a vitamin C deficiency disease that was rampant among the first European immigrants to the New World. Many a spindly scorbutic wretch was snatched from the Grim Reaper's clutch by a humble mug of spruce tea, whose recipe was passed to whites by different tribes of First Peoples.

One recipe is given here with the caution that you make certain which fir twigs you have collected. Yew, for example, is poisonous. Better not brew yew and it'll do you. Also, quaff not quarts of spruce or any other evergreen tea. Try a modest cup for interest. If you have scurvy (unlikely), see a doctor and chew vitamin C tablets on the way to the appointment. Don't pick White Spruce twigs to make tea. Its resin is too overpowering. If you are botanizing in eastern British Columbia or western Alberta, avoid Engelmann spruce. Its resin really stinks. Try Black Spruce, while remembering that some other conifers make pleasanter-smelling teas. Maritimers claim their native Red Spruce, *Picea rubens*, makes the most palatable infusion.

Recipe for Spruce Tea
Collect twig tips of fresh, young growth in late spring. Put a fistful of twigs in a warm pot. Fill pot with water heated to a rolling boil. Steep about five minutes. Sweeten with honey or maple syrup. Spice with a dust of cinnamon, a zest of orange peel, a nail or two of cloves, or bob dried blueberries in this fragrant tea.

In olden days, molasses was added to the steeped tea and it was left for days to ferment into a spruce beer with a very high vitamin C content. This type of tea was the famous antiscorbutic brew that saved Cartier and his men, and many other intrepid adventurers from death by scurvy. But different aboriginal peoples used different fir trees. For example, Cartier and his men were rescued by an infusion made from twigs of Eastern White Cedar.

Quotation
"In some of the new towns a liquor is made of spruce twigs boiled in maple sap."
Jeremy Belknap, *The History of New Hampshire*, 1792

SUMAC OR SUMACH

Genus: Rhus < *rhus* Latin < *rhous* Greek, an ancient name for Sicilian sumac, *Rhus corioria*, still featured prominently in Middle Eastern cookery where the berries are soaked to make a sour, fruity juice used to flavour yoghurt, sauces, and marinades. Don't try this with any North American species! The Greek *rhous* contains Indo-European **reudh* reddish, referring to the sumac's red autumn leaves.

Family: Anacardiaceae, the cashew family < *anacardium* Scientific Greek < *kardia* Greek, heart. Linnaeus borrowed the name to label the cashew from another earlier species that had heart-shaped leaves. The cashew family also encompasses mangos and pistachios but is home to poison ivy, poison sumac, and poison oak, all members of the *Rhus* genus.

French: *Sumac*

WORD LORE
Sumac or sumach < *sumac* Old French < *summaq* Arabic, the name for *Rhus coriaria*, part of the *materia medica* of Arab apothecaries, probably containing the Semitic root *smk* 'thick' in reference to the dense fur of small hairs that cover the branches.

SPECIES
Staghorn Sumac, *Rhus typhina* (Botanical Latin, reed-like < *typhe* Greek, reed, bulrush, cat-tail). Some botanist thought the cone-shaped fruit clusters of sumac resembled the brown female seed heads of cat-tails. Colonists called this species dyer's sumac from the yellow dye obtained from it. In the eastern United States, an old common name was Indian Lemonade, made by steeping the acidic fruit in water and adding a sweetener like honey. If you try it, be careful not to use the fruit of poison sumac.

Poison Sumac, *Rhus vernix* (Latin, varnish, because plant classifiers mistakenly thought that Poison Sumac yielded Japanese lacquer). Poison sumac is a small, uncommon tree in southern Ontario and southern Québec. Sap from most parts of this tree causes a nasty, irritating, often recurrent skin inflammation. Also in the same family of Anacardiaceae are small shrubs and vines that are poison, namely, poison ivy of eastern Canada, *Rhus radicans*, and western poison oak of British Columbia, *Rhus diversiloba*. Both these smaller nasties like to clamber up tree trunks where, if not recognized, they can be touched. Severe cases of poisoning, including eating any of these plants, has killed people. Therefore every wanderer in the Canadian woods should learn to identify poison ivy, poison oak, and poison sumac. A good start is: "Leaves of three, let it be."

Smooth Sumac, *Rhus glabra* (Latin, smooth, because it has no hair on twigs or leaves unlike other species). Smooth sumac is the most widely ranging species in Canada, found from Ontario to the interior of British Columbia. Smooth Sumac leaves were widely used in the nineteenth century in eastern North America to stop bleeding hemorrhoids.

CAUTION: POISON

Uses

For thousands of years, sumac has yielded a famous black dye. Goatskin was tanned black by sewing up the skins and filling the bag with a warm solution of sumac black. To dye the outside of the skins, the bags were inflated and floated in a vat of sumac black. Black Moroccan leather often owed its deep midnight hue to sumac dye.

❧

SYCAMORE

Genus: Platanus < *platanos* Greek, name for the eastern plane tree, *Platanus orientalis*. Like all sycamores, this plane tree had a very broad crown. *Platanos* is a contracted form of the earlier term *platanistos* which contains the Greek root **plat* 'flat, wide, broad' and one of the Greek superlative suffixes for adjectives, *-istos*, making the meaning 'the broadest-crowned' of trees, as indeed it was, both then and now.

Family: Platanaceae, the plane tree family

French: *sycomore, la platane*

WORD LORE

The word *sycamore* in one form or another has existed for more than three thousand years and has been used to name six or seven utterly unrelated trees. The oldest form is Hebrew *shiqmah*, a fig tree native to Egypt, Syria, and ancient Palestine, now called by botany *Ficus sycomorus*. The leaves of the *shiqmah* resemble those of the mulberry. Koine Greek, the Greek in which the New Testament was written, heard the Hebrew word as *sycomoros*, and a neat fit it was for a Greek ear with *sykon* the Greek word for fig and *moron* the Greek word for mulberry. Once formed, *sycomoros* took the usual route through Latin *sycomorus* into Old French *sicamor* to modern French *sycomore* and English *sycamore*.

The most plain-spoken and down-to-earth of the Old Testament's minor prophets was Amos who defines himself, in Amos 7:14 in this manner: "I was no prophet, neither was I a prophet's son; but I was a herdman, and a gatherer of sycamore fruit." Take that, Amos seems to say, a fig for all you spewers of ecstatic utterance, all you prophetic ravers. These are the visions of a common fig-picker. Actually, the Hebrew says "scraper," referring to scoring the bark of fig trees to induce better fruiting. But the metaphoric intent of the passage is clear. Amos was sent into exile back to his native land of Judah, for daring to speak against the authority of King Jeroboam. In the New Testament, in Luke 19:3-4, a short tax collector named Zacchaeus can't see Jesus passing because of the crowd "and he ran before, and climbed up into a sycamore tree to see him." *Ficus sycomorus* was planted as a shade tree beside roadways in the Middle East.

SPECIES

Sycamore, Plane Tree, *Platanus occidentalis* (Latin, western) ends its natural range in southern Ontario and is the only sycamore native to Canada. A hybrid called London Plane, *Platanus acerifolia* (Botanical Latin, maple-leaved) is sometimes planted as a shade tree on the streets of Ontario towns.

Both these sycamores have the typical spreading branches that form a broad, open crown and good height, averaging 35 metres, that make them such splendid shade-givers. Once upon a time butchers' blocks were sawn directly from sycamore trunks. In England, sycamore can refer to a big maple used as a shade tree, *Acer pseudoplatanus* 'false plane tree maple.'

Musical Reference

Handel's oft-played *Largo* is an arrangement of the much superior aria for soprano, *"Ombra mai fù,"* from his opera *Serse*, about the Persian King Xerxes. Musicological nitpickers who detest the arrangement as a profanation always begin by pointing out that Handel himself marked the passage "Larghetto." Our business with it here is simple. One of the sweetest melodies ever written on English soil, *"Ombra mai fù"* is an ode to a sycamore or plane tree, in praise of the tree's shade.

∾

TAMARACK OR LARCH

Other common name: Hackmatack

Genus: Larix < *larix, laricis* Latin < *larix* Greek, the tough-wooded European larch

Family: Pinaceae, the pine family

Québécois French: *tamarac*; Botanical French: *le mélèze laricin*

SPECIES

Tamarack, *Larix laricina* (Botanical Latin, larch-like, suggesting that the tamarack is the larchiest of larches) is also called hackmatack. The larches are among the few conifers that drop their needles every fall. A lover of peaty bogs and squelchy swamps, it grows from the Yukon to Newfoundland. But tamarack will also occupy open, airy sites that are drier than the acidic fens with full light that its seedlings prefer. In *The Deerslayer*, American novelist James Fenimore Cooper shows one character's knowledge of nature by having him state that "the tamarack is healthiest in the swamp."

WORD LORE

Larch

Larch was introduced into English in 1548, copied from the German *Lärch* by William Turner (1508-1568). Called "the father of English botany," this Protestant divine wrote some of the first works about plants based on his own research.

Tamarack

The Abenaki people live in Québec and Maine. They speak an Algonkian language in which their name is *Wabanakiyak* 'people of the dawn' or 'men of the east.' Canadian English borrowed tamarack from Abnaki. Tamarack, our common larch, stems from the fact that aboriginal peoples often named trees after the use made of their wood. Larch wood could be soaked in warm water and bent into gentle curves. So in the language of the Abenaki *akemantak* means 'wood for snowshoes.' We mangled the word into pioneer English as both hackmatack and tamarack. And snowshoes can still be made of the beautiful tamarack.

Uses

First Peoples used the thin roots of tamarack to sew together sheets of birchbark. Thus a character in Longfellow's *Hiawatha* invokes a great tree with these words: "Give me of your roots, O Tamarack!" As lumber, tamarack makes frames, poles, and pulp.

SURNAMES

Because the word entered English so late—in 1548 long after most surnames were established—Larcher and Larch are rare last names in English, and often are not the tree, but a contraction of a Norman French occupation name like *l'archer* 'the bowman.' Swiss German has surnames that denote an ancestor who lived beside a larch wood, namely, Larcher, Lerchner, and Lerchbauer.

Seeing My First Tamaracks

My first and most memorable site of tamaracks occurred on a raft trip I took in the 1970s down the Kootenay River in southeastern British Columbia. Five of us, including a professional rafter, were on a four-day jaunt to record a CBC radio program. Judy LaMarsh was the interviewer. Needless to say, that excursion was a lively one. Although the feisty Miss LaMarsh was not an accomplished outdoorswoman, her natural curiosity was great, and we had a professional forester on board to guide us. The day in question dawned as one of those brisk, spine-stiffening, scalp-corrugating, early fall days when autumn first clutches the soul and expands the spirit. Parts of the Kootenay have mild patches of rapids. Soon after breakfast we were on the water as the raft sported like a rubber dolphin around sassy whitewater bends and then moved more calmly as we entered a stretch where the river flowed smooth as a mirror, letting us observe the steep banks and beyond. On one side the fantastical pillars of hoodoos spiked into the sky. But on a high distant mountainside, lean splashes of amber yellow spoke sharply from among the other conifers. They were specimens of Western Larch, *Larix occidentalis*, whose autumn needles stood out yellow against the green pines beside them. Those tama-racks said loudly, "Autumn has come to the mountains." I have never forgotten the beauty of that announcement.

∿

Genus: Juglans

WALNUT & BUTTERNUT

Family: Juglandaceae, the walnut family, includes hickories, pecans, butternuts, and walnuts, about fifty species spanning the globe in the north temperate and tropical zones. From *iuglans, iuglandis* Latin, a walnut tree sacred to Jupiter, hence the name, a contraction of *Iovis* Latin, of Jupiter + *glans, glandis* Latin, acorn, any hard nut. In ancient Greece, one of the walnut trees was sacred to Zeus and the Romans simply translated the Greek phrase *Dios balanos* 'nut of Zeus' into Latin.

WORD LORE
Walnut was *wealh-hnutu* in Old English, literally 'foreign nut' or 'Celtic nut.' *Wealh* meant stranger, not of Saxon origin, hence the insulting origin of the names Wales, land of the *Wealas*, the Celts or Britons who were not Anglo-Saxon. The adjective of *wealh* was *welisc*, which gives modern English, *Welsh*.

A Juglans species with fruiting bodies

SPECIES

Two of the world's fifteen species of walnuts are native to eastern Canada. The roots of mature walnut trees produce a toxin that inhibits growth of other plants and saplings under its shade.

Butternut, White Walnut, *Juglans cinerea* (Latin, ash-coloured or grey, referring to the bark) grows from southern Ontario through to New Brunswick. Its soft wood is sometimes used in cabinet work and kitchen ware. The common name labels the light brown nuts. One of the nicknames for some Confederate soldiers during the American Civil War was "Butternuts" because a brown dye extracted from butternuts was widely used to colour homespun overalls worn by some Confederates when the army could not supply uniforms. Butternuts—actually the seeds—are oily and spicy.

Black Walnut, *Juglans nigra* (Latin, black, referring to the dark bark) occurs sporadically through southern Ontario. The wood, whose supply in Canada is gone, is a valuable lumber used in furniture veneers. In the United States, the rich, oily nuts are prized and harvested commercially.

The English Walnut, *Juglans regia*, is the source of walnut wood for furniture and of the commercial crop of walnuts used as food.

Quotation

"There are three sorts of walnuts; the hard, the soft, and another with thin bark. The hard sort bear a small nut, very good to eat, but apt to occasion costiveness, the wood of which is only fit to burn. The tender bears a large fruit, with a hard shell, the kernels of which are excellent. The wood of this tree is singularly curious, being almost incorruptible in water or in the ground, and difficult to consume in the fire: of this wood the Canadians make their coffins."
Thomas Anburey, *Travels Through The Interior Parts of America, 1771-1781*, first published in England in 1789.

WILLOW

Genus: Salix < *salix, salicis* Latin, willow tree

Family: Salicaceae, the willow family

French: *le saule* < **salha* Old French < *saus* Old French < *salix, salicis* Latin, willow

WORD LORE

Willow in its etymology is the tree whose leaves roll and turn in the wind. In Middle English it was *wilwe* from Old English *wilig* akin to modern Dutch *wilg*. Distant cognates include the Greek word for willow, *helike*, and the Latin verb *voluere* 'to roll, to cause to roll, to turn' and the Old High German *wellan* 'to roll.'

Salix

The Latin word is akin to the English sallow, pale-yellow, in reference to the leaves or bark of several European species. In Middle English, a *sallow* was one of the broad-leaved willows. Golden Willow, a *Salix alba* variety, has bright yellow twigs in winter.

SPECIES

Most of the more than 250 species of Salicaceae grow in the northern hemisphere, in a variety of forms, all the way from tiny shrubs like our dwarf Arctic Willow to imported weeping willow trees. More than 75 willow species are native to Canada, of which 8 are tree-size.

The smallest tree, often shrubby, is **Pussy Willow,** *Salix discolor* (Latin, of two colours, usually quite different; *discolor* does not mean 'discoloured') whose female catkins are collected for early spring bouquets, and can also be cut in late winter and brought indoors to force in a warm room.

Black Willow, *Salix nigra* (Latin, black, referring to the deep purple bark), is common from Ontario to the Atlantic, and is our largest willow. Polo balls are traditionally made from Black Willow wood.

In England, *Salix caerulea* (Latin, sky-blue, said of the bluish leaves) bears a common name derived from one sporting use, *Cricket Bat Willow.*

Uses

The bark of most willows (Latin *salix, salicis*) contains the antirheumatic glycoside, salicin. Willow bark was imported into ancient Egypt to treat fevers and aches. More than twenty-four hundred years ago Greek physicians prescribed extract of white willow for gout. Many of the First Peoples of North America chewed its bark to alleviate toothache. The Montagnais of eastern Canada brewed a rich tea of arctic willow leaves and drank it to reduce headaches. The active ingredient, salicin, breaks down in the stomach into salicylic acid. In the 1840s European chemists synthesized salicylic acid and it became a widely used antirheumatic. Doctors and patients noticed that it also brought quick relief from fevers, aches, and smaller pains of the body. Unfortunate side effects did include nausea and general gastric discomfort.

Then in 1899 the Bayer Company, a German pharmaceutical firm, instructed its chemists to try to find a synthetic derivative of salicylic acid that might not have such unpleasant side effects. They found a previously synthesized one called acetylsalicylic acid and began to market it under the name Aspirin, the 'a' for the acetyl radical, and 'spirin' because salicylic acid had been first isolated from spirin found in certain species of the garden shrub *Spiraea*. A long trip indeed, from willow bark to Aspirin, the most used synthetic drug ever created. New synthetic drugs have and will take over some of Aspirin's functions, but it still packs a healing wallop as an anti-inflammatory, analgesic, and antipyretic (reducing fever).

SURNAMES

Long associated with the University of Toronto, Canadian composer of church and organ music Healy Willan (1880-1968) has a surname that is a variant of an Old English dative plural of willow, *willen*, with the preposition lost, to indicate an ancestor who dwelt "at the willows."

Willoughby combines Old English and the Old Scandinavian suffix *-by* to render 'farm or settlement near willows.' Wilby though is Old English *wileg + beag* 'willow circle.' A crossing-place near water-loving willows gives Wilford. Willey is the name of half-a-dozen places in England, most meaning willow-lea, that is, clearing in the willows, from Old English *wilig + leah*. Wilton is a farmstead or village where willows grow.

French surnames in willow include Saule, Saulière, Saulais, Sauley, Saulet, and Sauleau.

Quotation

Weeping Willow, *Salix babylonica* (Medieval Latin, Babylonian), is a widely introduced species in Canada. The specific was named in error when Linnaeus thought the weeping willow native to southwest Asia. It came to Europe from the Far East. The eeriest psalm in the Old Testament is 137. It begins as praise of Jewish constancy during their Babylonian captivity and ends with a raucous incitement to infanticide. "By the rivers of Babylon, there we sat down, yea, we wept, when we remembered Zion. We hanged our harps upon the willows in the midst thereof." It seems the trees were actually *Populus euphratica*, Euphrates Poplar. Now to the conclusion of the psalm: "O daughter of Babylon, who art to be destroyed; happy shall he be, that rewardeth thee as thou hast served us. Happy shall he be, that taketh and dasheth thy little ones against the stones." How restorative and pious it is, to give voice to deep religious feeling in song.

∾

YEW

Genus: Taxus < *taxus* Latin, the yew tree. Because of its poisonous seeds, the Romans thought of the yew as the shrubbery the dead might encounter in hell. Yew had symbolic clout at Roman funerals. Ancient Greek commentators said the word was Latin. Yet Pliny has *taxicum venenum* 'yew poison' which looks mighty like a miswriting of *toxicon* Greek 'poison, arrow poison.' *Taxus* is certainly related to Greek, probably to *taxis* 'arrangement,' from the pleasing symmetry of yew needles in their tidy rows. Compare an adjective used by Theophrastos, the Greek writer about plants, who described one plant as *taxiphyllos* 'with leaves set neatly in rows.'

Family: Taxaceae, the yew family

French: *l'if*, yew wood is *bois d'if* and the famous Château D'If is Yew Castle.

WORD LORE
Yew gave the spelling heebie-jeebies to the Anglo-Saxons, so that we find Old English forms like *eow*, *iow*, *iw*, then in Middle English *ew* and *ewe*. Cognates of yew are widespread in the Indo-European languages and include Old Scandinavian *yr*, German *Eibe*, Welsh *yw*, Old Irish *ibar*, Gaelic *iubhar*, Old Slavic *iva* (semantically, willow), Gaulish *ivos* and hence Modern French *if*.

SPECIES
Western Yew, *Taxus brevifolia* (Botanical Latin, with short needles) is the only native Canadian species attaining tree size in British Columbia, and even it is usually a shrub. From

Manitoba east grows the low shrub, Canada Yew, *Taxus canadensis*. Archers like yew wood for the pliant bows that can be made from it. The five thousand-year-old Ice Man found in the Alps had beside his leathery remains a little axe handle made of yew, and a bow of yew wood.

Yews Are Poisonous

Taxine is a deadly alkaloid found in all species of yew, in the bark, leaves, and seeds. It is quickly absorbed once in the intestines. Survival after ingestion of quantities of yew seeds is uncommon. Children should be taught to avoid all parts of the plant. Do not burn yew clippings to ward off mosquitos and bugs as some dangerously foolish garden books suggest. Taxine can be carried in the smoke. Perhaps we could test it on the ignorant buffoons who whip off garden advice books over the weekend? Yew also kills livestock.

Yews Are Beneficial

Taxol, derived from Western Yew, may be of benefit as part of chemotherapeutic regimens in certain cancer treatments.

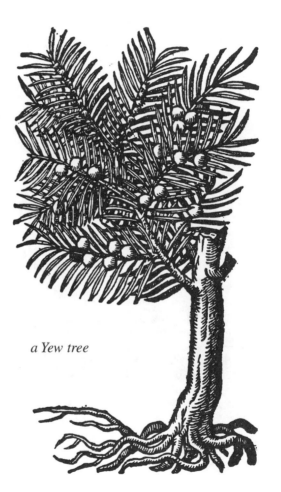

a Yew tree

WILD PLANTS

OF CANADA

WILD PLANTS OF CANADA

Here is a look at a few of the wildlings that have caught my eye and my etymological interest. There are tens of thousands of plants native to Canada. No book could contain them all. Note that some of this material has appeared in a less expanded form in a previous book, *Casselman's Canadian Words*.

From the crimson dye of *sang dragon* to flesh-devouring bog-dwellers, Canadian plants are beautiful, odd, and useful. Jack-in-the-Pulpit is very good for teenage acne. Prickly pear cactus leaves, despined, may be roasted in hot ashes, should you find yourself foodless in southern Alberta. The first French name for poison ivy was *l'herbe à la puce* or fleawort. But early French settlers said the name was far too mild.

Nature may well be the bounteous apothecary, but be warned: don't eat or apply internally or externally any wild plant unless you have correctly identified it and know its properties.

❧

CAUTION

Genus and species: *Salix arctica* < *salix* Latin, willow tree, akin to the English word *sallow*, pale-yellow, in reference to the leaves or bark of several European species. In Middle English, a *sallow* was one of the broad-leaved willows. There are dozens of species of ground-hugging dwarf willows in our north and they hybridize freely. Keeping low and clustered together is a protective strategy to keep out wind and freezing cold, and makes dwarf willows among the prostrate plant forms that grow in more elevated arctic regions.

ARCTIC WILLOW

Family: Salicaceae, the willow family < *salix* Latin, willow tree

SPECIES

Most of the more than 250 species of Salicaceae grow in the northern hemisphere, in a variety of forms, all the way from tiny shrubs like our dwarf Arctic Willow to imported weeping willow trees. More than 75 willow species are native to Canada, of which 8 are tree-size.

WORD & DRUG LORE OF SALIX

The bark of this scrawny, mat-forming shrub contains the anti-rheumatic glycoside, salicin. Willow bark was imported into ancient Egypt to treat fevers and aches. More than 2,400 years ago Greek physicians prescribed extract of white willow for gout. Many of the First Peoples of North America chewed its bark to alleviate toothache. The Montagnais of eastern Canada brewed a rich tea of arctic willow leaves and drank it to reduce headaches. See the *Willow* entry in the *Trees* section for the origin of aspirin from willow extracts.

∾

ARROWHEAD

Other common names: Duck potato, *flèche d'eau*, Tule, Wapata, Wapatoo

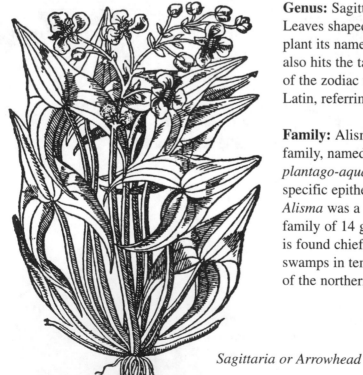

Genus: Sagittaria < *sagitta* Latin, arrow. Leaves shaped like an arrowhead give this plant its name. The Latin word for arrow also hits the target in Sagittarius, the sign of the zodiac that means 'the archer' in Latin, referring to the constellation.

Family: Alismataceae, the water plantain family, named after its type genus, *Alisma plantago-aquatica*, water plantain, as the specific epithet says in Botanical Latin. *Alisma* was a Greek water plant. This family of 14 genera and about 60 species is found chiefly in freshwater streams and swamps in temperate and tropical regions of the northern hemisphere.

Sagittaria or Arrowhead

SPECIES

The two common species in Canada are the more southerly *Sagittaria latifolia* with its *lata folia* or broad leaves, and the more northerly *Sagittaria cuneata* with cuneate or wedge-shaped leaves. Arrowhead tubers grow in the muddy guck of shallow streams and marshes across Canada, where wild geese, ducks, beavers, and muskrats chomp them with gusto.

Observing the animals feasting on tubers, native peoples found wapatoo could provide good food even in the winter. Adult aboriginal people used digging sticks to harvest arrowhead tubers, but children jumped into the streams and found tubers by squishing them between toes in the warm muck and yanking them loose. Wapatoo was then boiled or roasted in hot ashes. Wapatoo is Chinook Jargon, borrowed from an Algonkian language where *wap* or *wab* is the root for 'white.' Wapatoo means 'white food.'

BEECHDROPS

Genus: *Epifagus*, Botanical Latin < *epi* Greek, on + *fagus* Latin, beech tree, named because this plant grows parasitically on beech tree roots.

Family: Orobanchaceae, the broomrape family (Don't ask. They live at the other end of town.) The family is named after its type genus Orobanche < *orobos* Greek, a fodder plant, bitter vetch + *agchein* Greek, to choke, to throttle, because this broomrape grows parasitically on bitter vetch.

PLANT LORE

Beechdrops is a common parasitic plant of Ontario, and any-where else beech trees thrive. With no chlorophyl of its own, beechdrops absorbs food from the roots of beech trees, hence its botanical name. This dun-coloured and low-growing plant is often overlooked by the botanizing wanderer through a stand of beeches, even when its pallid flowers open in late summer. Beechdrops Pond is an official place name in Algonquin Park, Ontario. The term itself is not of Canadian origin but was coined by early settlers in Virginia who were clearing beech copses and noticed the flowers.

CANADA BLOODROOT

Other common names: Dragon's Blood, Puccoon, Tetterwort

Genus: *Sanguinaria*, Botanical Latin < *sanguis* Latin, blood. Its name in Algonkian languages is *poughkone*, often Englished as puccoon. The French is *sang dragon*, dragon's blood. All its names refer to the red dye that could be extracted by making a powder of the dried root of this member of the poppy family native to eastern North America. White pioneers learned from First Peoples to make a red dye for clothing and woven baskets. Puccoon also produced a ceremonial face paint.

Family: Papaveraceae, the poppy family, a large group of annual, biennial, and perennial herbs < *papaver*, Latin, poppy. The word was borrowed into Old English as *popig* and by Middle English was *popi*. Both *opion*, the Greek word for poppy juice, and *opos*, Greek, vegetable juice, as well as *papaver*, seem related to a Mediterranean root with a reflex in ancient Egyptian hieroglyphics as *peqer* 'poppy seed.'

SPECIES

Sanguinaria canadensis blooms late in April or early in May, with a single, short-lived, waxy, white flower, and does best in the rich soil of shaded woods. By midsummer the plant has completely died down. There is a bloodroot with a double flower that can make an attractive addition to the shade garden, if care is taken with soil preparation.

By the very earliest English visitors to North America bloodroot was called tetterwort. Tetter is an Old English word for any of various skin diseases like ringworm, impetigo, and eczema. Extract of Sanguinaria was used in the eighteenth and nineteenth centuries to treat warts and nasal polyps. A British doctor, J.W. Fell, read about the native peoples along the shores of Lake Superior who treated skin cancers with red sap of bloodroot and he tested it to his satisfaction in the 1850s.

In Russia, bloodroot is a folk remedy for skin diseases too. The Rappahannock people of eastern North America made tea from bloodroot as a specific against rheumatism. And other eastern tribes applied the crimson roots of Sanguinaria directly to decayed teeth as a remedy for toothache. Members of the poppy family contain many physiologically active substances that man may learn to extract and use. For example, biochemists have isolated an alkaloid called sanguinarine from bloodroot, but any therapeutic efficacy is so far much in dispute.

∾

Other common names: Convulsion Root, Fitroot, Fitsroot, Ghost flower, Indian pipe, Pinesap

Genus: *Monotropa* < *monos* Greek, single, solitary, alone + *tropos* Greek, a turning, a way of life, a manner, a direction. The gist of the name is 'living alone,' referring to the solitary, drooping flower on each stalk. These plants are saprophytes (*sapros* Greek, rotten, putrid, decayed + *phyton* Greek, plant); they grow on dead or decaying organic matter.

Family: Pyrolaceae, the wintergreen family, named after its type genus Pyrola < *pyrus* Latin, pear tree + *ola* Latin, diminutive ending, little, referring to the little leaves resembling those of a pear tree

CORPSE PLANT

SPECIES

Monotropa uniflora (Botanical Latin, with a single flower) is a pale saprophyte whose single, spooky-white flower nods downward and looks just like a pipe that a ghost might smoke. Without chlorophyll, this saprophyte gets nourishment from the fungus that grows on decaying humus strewn in the spectral gloom of dense conifer stands. In North American folk medicine of the early nineteenth century, the juice of Indian pipe stems used to be squeezed directly on inflamed eyes, just as the modern sufferer in search of a soothing eyewash might reach for a plastic bottle of Visine™. Folk medicine also had the notion that Indian Pipe would calm any of the various manias and fits that pioneers were liable to. Fitroot was often brewed up in the more obstreperous cases of cabin fever, for example, if Ol' Zeke down at Desolation Gulch just up and ate his chillun' one black February morning.

∾

DWARF DOGWOOD

Other common name: Bunchberry

Genus: Cornus < *cornus*, Latin, name for the cornelian cherry tree. Roman armourers prized the wood of cornelian cherry because of its hardness and used it to make the hafts of spears, javelins, and swords. Is this the origin of *cornus* with its similarity to *cornu* Latin, horn? Perhaps they thought the wood as hard as horn?

Family: Cornaceae, the dogwood family

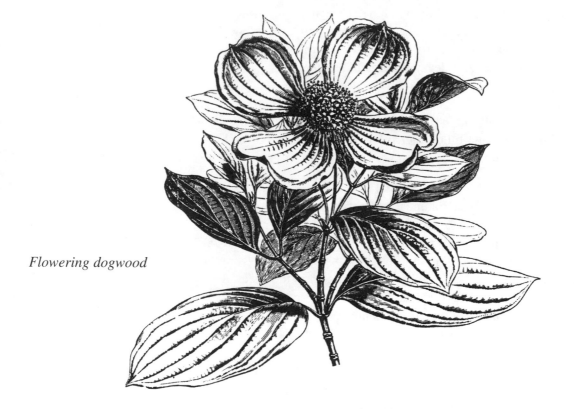

Flowering dogwood

SPECIES

Cornus canadensis or dwarf dogwood carpets moist, shady woodlands and logged-over areas all across Canada. In the autumn its leaves turn a deep bronzed plum colour, and it fruits in glistening clusters of orange-red drupes that result in dogwood's other common name, bunchberry. The prim symmetry

of massed dwarf dogwood makes it one of Canada's most
attractive wild flowers. An involucre of creamy-white bracts
surrounds each flowerhead like the starched and fluted ruff of
an Elizabethan collar, giving the little plant the air of a modest
courtier awaiting regal attention in some verdant chamber of
nature.

༄

Other common names: Blooming Sally, Mooseweed,
Willow Herb

FIREWEED

Genus: Epilobium, Botanical Latin < *epi* Greek, on + *lobon*
Greek, pod, refers to the placement of the flower in this genus
and the way it sits on top of a long ovary, which later becomes a
thin seed pod or capsule.

Family: Onagraceae, the Evening Primrose family < Onagra,
former botanical name of the family < *onagra* Greek, a female
wild ass < *onagros* Greek, male wild ass = *onos* Greek, donkey,
ass + *agrios* Greek, wild; but why the name? Because some
wandering botanist of yore saw wild asses eating evening
primroses?

SPECIES

August 17 is Discovery Day in Yukon Territory, recalling the
big strike of 1896 that began the Klondike Gold Rush. A sprig
of fireweed often decorates posters and advertising concerning
Discovery Day because the pink-blossomed fireweed, *Epilobium
angustifolium* is the official flower of the Yukon. When the pod
dehisces after midsummer, it sends out delicate aerial flotillas of
silky-winged seeds across thousands of northern acres. The
specific adjective *angustifolium* means 'with narrow leaves.'

Uses

Native peoples of the North liked the sugary pith of fireweed
obtained by splitting a young stalk and scooping it out. Elk and
deer browse in fireweed fields, and the dark, sweet-smelling
honey makes it worth putting beehives near fireweed.
Beekeepers also plant fireweed close to apiaries because the
honey produced is of superior taste. French-Canadian voyageurs

called it *l'herbe fret* and cooked it as greens. In Russia, fireweed leaves are brewed for kapporie or kapor tea.

When invasive as on cleared or logged-off land, it sometimes bears the name mooseweed. Its most common name signifies that fireweed is among the first plants to bloom on land after a burn-over. Campers in our North may brew up a refreshing backwoods tea by pouring hot water over the tender young leaves of fireweed. The tea is light green and sweetish. Just make sure you have correctly identified fireweed before teatime, so that there may be an after-teatime.

~

GROUNDNUT

Mi'kmaq name: *Sequbbun*

Other common names: Bog potato, Micmac Potato, Travelers' Delight

Genus: Apios < *apios* Greek a pair, because the tubers on an individual rhizome seem to grow in pairs

Family: Leguminosae, the legume or pea family, the second largest family of plants with more than 550 genera and at least 13,000 species < *leguminosus* Latin = *legumen* Latin, what one gathers or picks, notably peas and beans + *osus* Latin adjectival suffix indicating 'abounding in, full of.'

SPECIES

Groundnut, *Apios americana*, has dark-red or brown flowers that resemble those of sweet pea. They thrive in damp ground from Nova Scotia to Ontario. Mi'kmaq people prized the sweet tubers of this plant, which has a chestnut flavour, and they call it *sequbbun*. Early white settlers in the Maritimes shared the taste. The Nova Scotia town and river, Shubenacadie, is a French and English attempt at a Mi'kmaq phrase that means 'sequbbun (groundnuts) grow here.' One healthy plant may have ten or twelve tubers.

~

Genus: Hepatica < *hepar* Latin, liver, referring to the mottled, liverish colour and the shape of the leaves that come only after the pale purple flowers have faded

Family: Ranunculaceae, the buttercup family, named after its typical species, the buttercup, Ranunculus < *rana* Latin, frog + *unculus* Latin, diminutive suffix, little, tiny. *Ranunculus* is Latin for little frog, so named because buttercups like damp places just as amphibians do.

HEPATICA

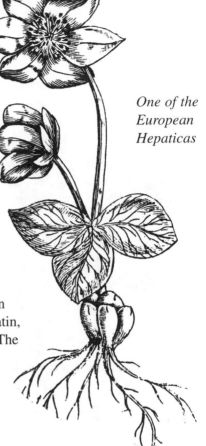

One of the European Hepaticas

SPECIES

No better introduction to this early little beauty exists than the entry by our pioneer writer Catharine Parr Traill. In *The Backwoods of Canada* published in 1836 she wrote, "The hepatica is the first flower of Canadian spring: it gladdens us with its tints of azure, pink, and white, early in April, soon after the snows have melted from the earth. The Canadians call it snow-flower, from its coming so soon after the snow disappears." Two species brighten eastern Canadian woodlands, *Hepatica acutifolia* (Botanical Latin, with sharply pointed leaves) and *Hepatica americana*. The earliest British settlers thought of it as—Noble Liverwort. Indeed.

Other common names: *Gouet à trois feuilles*, Indian Turnip

Genus: Arisaema, Botanical Latin < *aron* Greek, any arum plant + *haima* Greek, blood, because certain European species have reddish spots on the leaves

Family: Araceae, the arum family < *aron* Greek, name of an aroid plant related to our Jack-in-the-Pulpit. This family has 15 genera and more than 2,000

JACK IN THE PULPIT

species, chiefly tropical. Most arums have arrow-shaped leaves and inconspicuous flowers.

SPECIES

Jack-in-the-pulpit, *Arisaema triphyllum* (Botanical Latin < *treis* Greek, three + *phyllon* Greek, leaf) the familiar little arum of our moist woods, refers in its specific name to the three little leaflets of which each leaf seems to be comprised.

French-Canadian lumberjacks encountered the plant in Québécois woodlands and thought the spathe that bends over the jack or spadix like a little pointed flap looked like one of their implements, a hand-held log-hook or *gouet*, hence the French name, *gouet à trois feuilles* 'hook with three leaves.' Iroquois peoples named it *kahahoosa* 'papoose cradle,' because it looked like the backrack Iroquois women used to carry their babies. Indian turnip was a pioneer name for the plant.

The corms of jack-in-the-pulpit can only be eaten well cooked. Consumed raw, they burn the tongue, mouth, and throat, and are highly poisonous. Scottish settlers called it Devil's Ear. Medical records contain case histories of those who ate jack-in-the-pulpit raw and died from a hideously painful gastroenteritis. Colonists in early Virginia dubbed it American wake-robin, but the name has no practicality as a folk term because it was used to refer to at least six separate plants, all of different plant families, including the trillium.

An Arisaema species

❧

Other common names: Bearberry, Bear's Grape, Indian Tobacco

Genus: Arctostaphylos < *arktos* Greek, bear + *staphylos* Greek, berry

Family: Ericaceae, the heath family < *erice* Latin, a heath plant < *ereike* Greek, heath plant < *eri-* Greek root meaning woolly, applied to many plants with downy leaves or stems

SPECIES

Kinnikinick means 'mixture' in Cree and Ojibwa, specifically a smoking mixture that might contain dried bearberry leaves, dried sumac leaves, red-osier dogwood bark, and tobacco. This very pungent Indian tobacco was smoked in a pipe.

The red fruit of the bearberry is much munched by bears too, hence its common name, and its botanical label: *Arctostaphylos uva-ursi. Arctostaphylos* means bearberry in Greek. *Uva ursi* means grapes of the bear in Latin. The combined English, Greek, and Latin of the botanical nomenclature appear to make clear the ursine connection, eh?

Grouse browse bearberries. Deer delight in it. Gardeners looking for a crisp groundcover ought to ponder its use. The tiny flowers look like white bells with pink rims, and the dense mats of leathery, evergreen leaves are most attractive.

KINNIKINICK

∾

Other once-common name: Wishakapucka

Genus: Ledum, Botanical Latin < *ledon* Greek, the name for a European evergreen shrub we generally call Rock Rose, *Cistus*

Family: Ericaceae, the heath family

SPECIES

Labrador tea, *Ledum groenlandicum* (Botanical Latin, of Greenland), an evergreen shrub of the heath family, has been a staple infusion of northern peoples since the first humans

LABRADOR TEA

crossed from Asia to America some 12,000 to 40,000 years ago, give or take a day. One Ledum species is circumpolar, so anthropologists posit that First Peoples may have brought knowledge of its refreshing and medicinal properties with them. The tea is made by lightly steeping cleaned, crushed, dried leaves. Arctic explorer Sir John Franklin in his 1823 *Narrative of a Journey* reported that the tea smelled like rhubarb. It acted as a mild digestive and perked up one's appetite. The Hudson's Bay Company for a time imported the leaves into England where they enjoyed popularity under the peppy name of Weesukapuka! Canadians spelled it in a variety of ways, usually referring to the plant as wishakapucka, their attempt at the Cree term, *wesukipukosu* 'bitter herbs.' The leaves might also be added to kinnikinick as part of the standard native smoking mixture.

~

MAYAPPLE

Genus: Podophyllum, Botanical Latin, shortened from the original name in old botany, Anapodophyllum or duckfoot leaf < *anas* Greek, duck + *podos* Greek, foot + *phyllon* Greek, leaf. The basal leaf does have a long, foot-like stem. This genus includes about 450 species of shrubs.

Family: Berberidaceae, the barberry family, perennial herbs and shrubs in about twelve genera, mostly of the northern temperate region < *berberis* Old French, one of the barberries < *barbaris* Latin, one of the shrubs < ultimately *Barbaroi* Greek, literally 'stammerers,' but any foreigner who could not speak Greek was a *barbaros* and was thus uttering nonsense syllables like 'bar-bar.' This notion is of ancient Indo-European provenance. Consider *barbaras* Sanskrit, stammering. A barberry-like shrub must have sometime come from North Africa or the Middle East. Note that Arabic borrowed the Greek term too, as *al-Barbar*, originally any people the Moors encountered who could not speak Arabic, notably the Berber people, whose name reflects this Arab prejudice. And English still refers to barbarians.

SPECIES

Mayapple possesses the pleasingly plosive appellation of *Podophyllum peltatum. Peltatus* is classical military Latin for 'armed with a pelta.' The pelta was the sturdy little shield, shaped like a half moon, carried by the Roman infantry whose armourers borrowed it from the *pelte* shield of the ancient Thracians. Each of the mayapple's two large leaves resemble such a shield.

Uses & Cautions

The leaves and roots are poisonous. The ovoid yellow fruit of the mayapple is edible only when it is ripe. Before it ripens, the fruit may be quite dangerous to consume. Like many common names of plants, mayapple is a slight misnomer, since the yellow fruit does not appear until later in the summer. But, when it did ripen enticingly, the mayapple sent pioneers and even visiting explorers into dizzy dithyrambs of praise. Here is W. Ross King, author of *The Sportsman and Naturalist in Canada* (1866): "a delicious and refreshing wild fruit … about the size of a bantam's egg … presents a mass of juicy pulp and seeds, not unlike pineapple in flavour." Steady on, old man. Ripe fruit can be done up as preserves and added to jams and jellies too. Many moderns who have tasted it say mayapple berries are bland, acidic, and too pulpy.

While there are many dangerous folk remedies connected with extracts of mayapple, like its unadvised use as a purgative, pharmaceutical investigation has led to the clinical use of podophyllin, a resin from the mayapple, and its less toxic derivatives like epipodophyllotoxin. In 1977 podophyllin was the drug of choice in the treatment of venereal warts or condylomata acuminata. American medical literature reports that the Penobscot peoples of Maine treated certain cancers with an extract from the rhizome of the mayapple. And so the search goes on.

CAUTION: POISON

PICKEREL WEED

Genus: Pontederia, Botanical Latin,named to commemorate Giulio Pontedera, an eighteenth-century professor of botany at the University of Padua

Family: Pontederiaceae, the pickerel weed family

SPECIES

Many an angler of eastern Canada has squatted in a boat patiently hoping that a pickerel weed marsh in the calm summer waters of a shallow cove will live up to its name—soon! Canada's sweetest eating fish does frequent stands of pickerel weed, *Pontederia cordata*, with rootstocks that creep through the mud, one large heart-shaped (Latin, *cordatus*) leaf, and many blue florets on a big phallic flowerspike that pokes into bloom late in summer. Pickerel lay their eggs on the submerged stalks of the plant. Owners of aquatic and bog gardens use pickerel weed as an ornamental, although the blue flowers are quite short-lived. It's a relative of the much more widely grown aquatic ornamental, water hyacinth. In the American south the plant is called wampee, and in Florida, it's alligator wampee, for gators love to lurk and slumber in hideaways of disguising foliage that *Pontederia* provides.

❧

PIPSISSEWA

Genus: Chimaphila, Botanical Latin < *cheima* Greek, winter + *philos* Greek, loving, winter-loving because the leaves stay green all the year round

Family: Pyrolaceae, the wintergreen family, named after its type genus Pyrola < *pyrus* Latin, pear tree + *ola* Latin, diminutive ending, little, referring to the little leaves resembling those of a pear tree

SPECIES

If you've ever taken a swig of good, home made, tongue-startling, palate-corrugating root beer (not the homogenized, limp-bubbled suds of commercial root beers), then you know the refreshing, wintergreen-like taste of Pipsissewa. The word comes from the language of a people inhabiting northeastern Canada, the Abenaki. In their language, Abnaki, *kpi-pskw-àhsawe* means 'flower of the woods.' Its name in botany is *Chimaphila umbellata*. The specific refers to the umbel or loose terminal cluster of little waxy heath-like flowers of pink hue that bloom in the summertime in dry evergreen woodlands. Nowadays oil of wintergreen is synthesized, but its chief active ingredient, methyl salicylate, is found in Pipsissewa leaves. Outdoorsy folk still chew the leaves for their minty brio, and the plant is widely used in herbal remedies.

∾

Other common names: Indian Cup, *Petits Cochons* 'piggywigs,' Whip-poor-will's boots

Genus: Sarracenia, Botanical Latin, commemo-rates Dr. Michel Sarrazin (1659-1734), Canada's first professional botanist. He came out to *La Nouvelle France* in 1685 to become surgeon-major to the colonial army, and later co-wrote one of our pioneering botanical works, *L'Histoire des plantes de Canada*.

Family: Sarraceniaceae, the pitcher plant family of bog-dwelling, insectivorous, perennial herbs

SPECIES

Sarracenia purpurea (Botanical Latin, purple, referring to the colour of the mottled pitchers) is the floral emblem of Newfoundland. Pitcher plant is the stout little carnivore of Canada's peat-quilted swamps and jelly-carthed bogs, where it traps insects in leaves modified to hold water, hence pitcher plant. The slippery sides of each pitcher are lined with downward-pointing hairs that help insects slide into the pitcher but prevent them from escaping. Trapped without mercy, they struggle, fall exhausted back into the water, and drown in

PITCHER PLANT

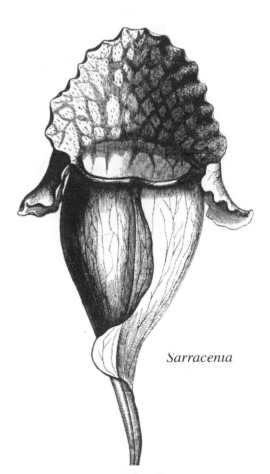

Sarracenia

the liquid to which the plant has added a flesh-dissolving enzyme. The decomposed bodies of the insects provide essential nutrients for the pitcher plant.

◆

PRICKLY PEAR CACTUS

Genus: *Opuntia* Botanical Latin < *Opos, Opuntis* Greek, name of a place in Locria, in ancient Greece, noted for some spiny plant, perhaps one that gave a milky latex like that from certain figs < *opos* Greek, milky juice of figs

Family: Cactaceae, the huge cactus family, made up of about 85 genera and some 2,000 species < *kaktos* Greek, the prickly cardoon plant, related to an artichoke, still cultivated for its edible root and petioles, and, incidentally, the flavouring in a revolting (to me) apéritif sold under the brand name of Cynara. Most dictionaries say the Greek word is of unknown origin. Uh-huh. Greek *kaktos* looks borrowed from ancient Egyptian *qatcha* 'thorns, thornbush.' The Egyptian term was heard by Greek ears as if it contained their word *kakos* which meant bad or injurious, in reference to the spines of the cardoon.

SPECIES

Opuntia polyacantha (Botanical Latin < *polys* Greek, many + *acanthos* Greek, thorn, spine) has indeed many spines. On the dry banks of old coulees in southern Saskatchewan and Alberta, on the dry hills of British Columbia's interior, even on some dry islands in Georgia Strait, the prickly pear cactus blooms in early summer. Surprisingly, it also grows on Pelee Island in Lake Erie. Among the cat's cradle of prickly spines sit sensual cups of translucent yellow sepals, seemingly spun of buttery silk. These cactus flowers really do look like sexual organs, and of course they are, as they beckon insects into their vulval bowls with the promise of nectar, and send them on their way with a freight of pollen, male microspores eager to egg-on any friendly ovum in the area. But sweet as the flowers are, prickly pear spines penetrate shoe leather quick as steel needles. They hide under snow too and stiletto straight through a ski boot. But if you skin the stubby leaves and carefully pluck out the spines, you can eat the leaves after roasting them in the bottom of a hot campfire.

◆

Genera: *Gyrophora* and *Umbilicaria*

ROCK TRIPE

SPECIES & USES & WORD LORE

Rock tripe is a translation of the Canadian French coinage *tripe de roche*, here meaning 'rock guts.' Native peoples first showed whitemen how to eat this emergency food, which they called *wakwund*. Voyageurs often scraped this edible lichen directly off the rocks from their canoes, and sometimes carved their initials in the blank rock wall so exposed. It is not highly nutritious but does fill the stomach of a starving wretch until he finds his fellows, his fate, or some real food.

Lichen is a symbiotic partnership between a fungus and an alga. The fungus supplies the outer form of the lichen, the alga supplies chlorophyll so photosynthesis can take place.

Here's what explorer Samuel Hearne said about rock tripe in 1795 in *A Journey from Prince of Wales's Fort:*

> There is a black, hard, crumply moss, that grows on the rocks … and sometimes furnishes the natives with a temporary subsistence, when no animal food can be procured. This moss, when boiled, turns to a gummy consistence, and is more clammy in the mouth than sago; it may, by adding either moss or water, be made to almost any consistence. It is so palatable, that all who taste it generally grow fond of it. It is remarkably good and pleasing when used to thicken any kind of broth, but it is generally most esteemed when boiled in fish-liquor."

Tripe started life at the back of the butcher shop. It's tissue from a cow's first or second stomach used as food. Tripe came into the English wordstock from Norman French after 1066 and all that. The French borrowed it from Provençal *tripa* and cow-stomach-eating troubadours heard it first in Italy as *trippa*. English extensions of the sense followed, and tripes meant guts, then tripe was a worthless person, food, or thing.

❧

ROSE POGONIA

Other common names: Adder's Mouth, Snakemouth, Snake's Tongue

Genus: Pogonia, Botanical Latin < *pogonion*, Greek diminutive, little beard < *pogon* Greek, beard, referring to the exquisite crests and fringes of its yellow, purple, and white labellum or lip.

The *pogon* root also gives rise to two obscure and silly English words, pogonophoric 'wearing a beard' and pogonotrophy 'the growing of a beard.'

Family: Orchidaceae, the orchid family < *orchis* Greek, human testicle, referring to the paired tubers of some native Greek orchid—because of which herbalists have for millennia considered some orchid root extracts to be aphrodisiacs.

SPECIES

Rose pogonia is botanically *Pogonia ophioglossoides* (Botanical Latin < *ophis* Greek, snake + *glossa* Greek, tongue + *oides*, similar to < *eides* Greek, form, shape). It is thus a translation of an old common name, Snake's Tongue. Rose pogonia is a startling pink orchid of peat bogs in eastern North America that may with difficulty be grown in a bog garden with a hefty pH of 4 or 5. Rose pogonia spreads through the squelchy sphagnum of bogs by means of underground rhizomes, and so an itinerant bog-trotter, bored with yet another stand of rotten stumps, may suddenly have his heart lifted by coming upon a mass of pogonia all in bloom on a hot summer afternoon. Even the miasma of a fetid swamp may be disbursed, as the subtle spice of pogonia fragrance dances in the air.

❧

SALMONBERRY

Other common name: Ollalie

Genus: Rubus, Botanical Latin < *rubus* Latin, red; and an old botanical name for any bramble bush. The Latin root pops up in the English word *ruby* and in learned adjectives like rubefacient 'causing redness.'

Family: Rosaceae, the rose family

SPECIES

Salmonberry, *Rubus spectabilis*, recently gave its name to a Canadian movie starring Alberta's k.d. Lang. In *Salmonberries* the frisky warbler of pop songs played a provocative but beguiling role. So too did the sweet red berries of this shrub that belongs to the huge rose family of plants. *Rubus* is a randy genus where species interbreed at the drop of a pollen grain, so that there are hundreds of named varieties. Flowers are rosey red. The specific, *spectabilis*, means showy. Its juicy fruits look like big raspberries and are eaten ripe or made into a delicious jam. The common name was used first along the banks of the Columbia River where native peoples had a favourite dish that consisted of the very young, tasty shoots of the plant eaten with dried salmon roe. Indeed, salmonberry's home range is the Pacific coast, and it thrives west of the Rockies where it was called ollalie in Chinook Jargon.

∾

Skunk cabbage comes in a western and in an eastern Canadian type, and they are swamp plants of similar habit but of entirely different genera.

Genus: *Lysichitum americanum*, Western Skunk Cabbage, takes its generic name from Greek where it means 'loosed chiton,' referring to the shedding of the giant spathe. Now the chiton was one ancient Greek equivalent of a tunic. It could be made of homespun cotton, of sturdier and more expensive linen, or, on a prostitute, it might be of the flimsiest gauze. If an Athenian citizen or a slave in the fifth-century city wore too loose a chiton, local prudes would *tsk-tsk* and declare it an outer sign of an inner moral laxity. Wearing a loose chiton meant the wearer could simply flip it up over his or her head for a quickie nooner in the boscage by the portico. After such a carnal connection, one nipped into any nearby temple of Athena for a thirty-second prayer to the virgin goddess, and, all sins absolved, went merrily on one's way, sandals flip-flopping on the cobbles.

Genus: *Symplocarpus foetidus*, Eastern Skunk Cabbage does stink, as its name makes plain. The Latin adjective *foetidus*

SKUNK CABBAGE

gives us fetid in English. The generic means 'with connected fruit' referring to the aggregrate fruit formed from joined ovaries.

Family: Araceae, the arum family < *aron* Greek, name of aroid plant related to our Jack-in-the-Pulpit

WESTERN SPECIES

In the thick-aired, sepulchral gloom of a rain forest bog one spies the bold yellow sword that is the spathe of Western Skunk Cabbage poking up as early as February, emerging in the coolth with its spadix coyly sheathed in a yellow tunic. This tunic or spathe later opens to reveal the flower-bearing rod of the spadix within. Some plant books label western skunk cabbage as "evil-smelling" and "with a foul stench." I lived in British Columbia for three years in the 1970s, most happily in a cottage on the Beach Grove Road near Tsawassen. One day in early spring I took the ferry to Vancouver Island, and went tromping and swamping for a day or two in Pacific Rim National Park. I certainly saw vast platoons of serried skunk cabbages in fenny glades a few yards from the Pacific waves. They did not smell skunky at all. The odour that *Lysichitum* gives off probably imitates the sex pheromones of several swamp insects that pollinate the plant. The aroma is one of fresh, primal fertility, of the vernal surge of life. Bears wolf down—so-to-speak—the whole plant, rootstock and all. The root can also be roasted, dried, and pounded into a good flour, much like its Polynesian relative, taro.

EASTERN SPECIES

Eastern Skunk Cabbage often is the first flower of spring in eastern Canadian wet places. The spathe that at first encloses the spadix is not yellow like the western version but a mottled brown-purple.

CAUTION

While we are discussing arums (members of Araceae), it is appropriate to mention a common toxin that plants of this family contain. Calcium oxalate accumulates in the fleshy parts of arums like their leaves and stems. The microscopic structure of a calcium oxalate crystal is nasty. One might say its crystal resembles a long needle made of minuscule razor blades. What happens when calcium oxalate crystals become embedded in mucous membranes that line the human mouth and throat? Pain, irritation, and swelling produced by the irritation, and then hysterical apprehension that the victim is choking; all are frequent symptoms of ingestion.

This habit of calcium oxalate gives a common name to a houseplant called dumbcane, found in many houses with small children and pets. Dumbcane ought to be banned from greenhouses and plant stores. It contains massive accretions of calcium oxalate. Fortunately for many peoples who depended on the roots of arums for flour-making food, a great deal of crystalline oxalate is dissolved when plant parts are boiled in hot water. Though not a Canadian plant, dumbcane or *Dieffenbachia seguine*, is a real pest. Perhaps I am hypersensitive because, as a small child, I watched a playmate gag and choke after chewing a big dumbcane leaf. I always wished the jerk who sold it to his mother had partaken of it.

❧

Other common names: Adam and Eve, Fawn Lily, Glacier Lily, Yellow Avalanche Lily

Genus: Erythronium, Botanical Latin < *erythros* Greek, red, referring to the one Eurasian species that has pink to purple flowers

Family: Liliaceae, the lily family < *lilium* Latin, lily

SPECIES

Erythronium grandiflorum (Botanical Latin, big-flowered) is the Snow Lily or giant Dogtooth's Violet of the Pacific mountains. It is frequently called Glacier Lily too, and took that name because of its abundance in Glacier National Park near Golden, British Columbia, and not too far from Banff. The plant has a plenitude of common names, including Adam and Eve, and in California, Chamise Lily. Glacier Lily may be called Fawn Lily, because its leaves are like dappled flanks of fawns. One older

SNOW LILY

English name is Adder's Tongue. A peppier North American nickname is Yellow Avalanche Lily, which at least is a mnemonic for this plant's big, clear-yellow flower that pokes through a shawl of melting snow on many a spring slope.

Uses
The bulbs of Snow Lily have in the past been dried and stored for winter food, but since the literature of veterinary science reports that bulbs of some Dogtooth's Violet species have poisoned chickens, don't be too eager to try this. Siksika people of the west used Snow Lily root to treat some skin disorders.

TRILLIUM

Other common names: Mooseflower, Wake-robin

Genus: Trillium. The genus has more than thirty species, all of them with three-part leaves and flowers, hence the first root in the botanical name. The great Swedish botanist and founder of systematic botanical taxonomy, Carolus Linnaeus (1707-1778), may have named the genus; and so the *Oxford English Dictionary* suggests that trillium is a Latinizing of the Swedish word for triplet, *trilling*. Maybe. There's no written proof of that. A member of the lily family, trillium might also be a shortening of *tri* and *lilium*, three-part lily. It is certainly not, as many American botanical books state, from *triplum*, Latin for triple.

Family: Liliaceae, the lily family < *lilium* Latin, lily

SPECIES
The great white trillium, *Trillium grandiflorum* (Botanical Latin, big-flowered), is the floral emblem of Ontario. A common colonial name was wake-robin, because it blooms in spring. But much more frequently, in earlier diaries and pioneer letters home, the trillium was simply called a lily. In Nova Scotia settlers called it Moose-flower, according to the 1868 *Canadian Wild Flowers* by Catharine Parr Traill. One species on our west coast is *Trillium ovatum*, or western wake-robin.

WATER LILY

Genus: Nymphaea < *Nymphe* Greek, a minor water goddess or sprite < *nymphe* Greek, a young, nubile maiden < *nyos* Greek, daughter-in-law

Family: Nymphaceae, the water lily family

SPECIES

Nymphaea odorata is the common white water lily of eastern North America. Many-petalled, sweetly odorous, and of exquisite symmetry, the flowers open early in the morning on the still waters of quiet ponds and waveless inlets where their fragrance attracts to the floating pads of their leaves damsel-flies and all manner of tiny fauna like toadlets and turtles. One race of semi-divine nymphs, the naiads, took particular joy in haunting brooks and silvan streams. Nymphs were the guardian spirits of fountains, rivulets, and creeks.

An African species of Nymphaea

White water lily spreads its seeds by means of a neat adaptation. Water lily seeds dwell in a capsule of air-filled, spongy tissue. The capsule breaks free of the lily pad and bobbles lazily downstream—sometimes for kilometres—before the spongy air spaces slowly fill with water so that the heavy capsule sinks to the bottom, settling into fertile muck to root. The seeds of some Eurasian water lilies were taken to quell the tumult of sexual desire, and thus were termed anaphrodisiacs.

Related Species

A very close relative of our white water lily is *Nymphaea lotus*, the sacred floral symbol of divinity in ancient Egypt and Greece. In sandy hieroglyphs reposing on a lotus pad shines Ra, sun god of the old Nile. Or Ra appears springing up from a lotus blossom. The ancients believed the lotus represented intellect, for the flower of human thought rises from the muddy clay of our mortal bodies, just as the water lily's beauty has its origin in the muck of a river bottom. Various plants in history have been called lotus. Hindu mythology's plant is a water lily of the genus *Nelumbo*. The same *Nelumbo* is the sacred lotus of China. The Lotus-Eaters of Homer may have munched on *Ziziphus lotus*, a North African shrub whose fruit was edible.

ॐ

WILD GINGER

Genus: Asarum, Latin < *asaron* Greek, a wild ginger

Family: Aristolochiaceae, the birthwort family, named after its type genus Aristolochia, a plant like Dutchman's Pipe, believed by the ancients to assist labour in childbirth < *aristos* Greek, best + *locheia* Greek, childbirth

SPECIES

Wild ginger, *Asarum canadense*, clings to the damp humus of shady forest hummocks. Its dense, kidney-shaped leaves usually hide the brownish-purple flower that grows at ground level to facilitate pollination by crawling insects. Wild ginger spreads also by a creeping rhizome.

Uses

First Peoples taught early white settlers to peel the root for use
as a spicy flavouring. To pep up pioneer baked goods, wild ginger root was boiled with sugar as a bread and pastry spice.
Many Algonquin tribes of eastern Canada made a wild ginger
tea to relieve jumpy heartbeat, although today cardiac arrhythmias
are best treated by a doctor. Other native North Americans
steeped the roots and poured the liquid into the ear to treat
minor earaches. A pioneer toothpaste of powdered black alder
and black oak bark was made palatable by adding an equal
portion of ground-up wild ginger root. On the west coast, a
Pacific species, *Asarum caudatum* (Botanical Latin, with a tail,
referring to the lobes of the calyx which are formed like a long
tail) provided a spring tonic tea when Skagits people and their
neighbours to the north boiled the leaves of wild ginger.

Many groups of first North Americans also used high
concentrations of the root extract as an emmenagogue, that is,
an agent to promote menstruation. As they did of many plants
so used, native healers also thought wild ginger extract could
induce abortions if given in sufficient strength and dosage.
Needless to say, none of these ancient remedies should be
taken—if ever—except under competent medical supervision.

Modern science has extracted from wild ginger, aristolochic
acid, which has some anti-microbial effect, and from the root, a
broad-spectrum bactericide of limited use in some prescription
cough medicines. A final caution about wild ginger root is that
some of the essential oils have caused cancerous tumours in lab-
oratory tests.

∾

Other common name: Silverberry

WOLF WILLOW

Genus: Elaeagnus < *elaion* Greek, olive oil, any oil + *hagnos*
Greek, holy. The holy oil was perhaps a sacred lotion extracted
once by unknown southern Europeans from a plant of the
willow family. Another guess is that the initial root in Elaeagnus
is *helodes* Greek, growing in marshy places, and, except for the
fact that even the species of the Far East do NOT grow in

marshes, it is certainly a very remarkable example to us all of etymological inquisitiveness.

Family: Elaeagnaceae (my, what a sonorous wordlet!), the silverberry family, and good luck to it!

SPECIES

Wolf willow, *Elaeagnus commutata*, is also called silverberry bush because of the silvery sheen of its leaves, flowers, and fruit. A former specific was *argentea* 'silvery' in Latin. Hardy all across our prairies, wolf willow likes especially the dry banks of rivers and streams. The greyish-silver colour of the shrub gave good camouflage to the prairie wolf, hence its common name. Silvery, yellow-lined flowers give off a sickly sweet and powerful aroma in early summer. This shrub makes a useful windbreak for a prairie field, and is easily grown in dry sites.

Uses
Native American women showed pioneer ladies how to make bead necklaces and bracelets from the hard brown nuts obtainable after boiling off the mealy flesh of the silverberry. Wolf willow nuts were also strung on buckskin clothes to decorate them and to weigh down the buckskin fringes so they would hang straight.

For an unsentimental glare at the hardships of homesteading in southern Saskatchewan, I do recommend a 1962 novel by Wallace Stegner called *Wolf Willow*. The author blends historical fact with what are clearly family memories as he displays how a frontier pickaxes the human soul chip by chip.

ENVOI

It is, I hope, fitting to finish off our romp with a moment of light raillery. We former colonials are always intrigued by the special uses to which the British sometimes put their gardens, uses that come to light in the pages of British tabloid newspapers.

BRITISH GARDEN GUIDE

MAY
Sharpen tools.

Plan to prune family tree.

Feed lime pit.

JUNE
Clear upstanding garden bones.

Cut back climbing aunt.

Drink plenty of lukewarm Bovril.

JULY
Poison Nigel's kipper.

Get Dad to inhale pesticide.

Disinter Mrs. Innis.

Bury Nigel near roses.

AUGUST
Feed cat to Dora.

Put Mrs. Innis under hydrangea.

Have a nice cup of tea.

SEPTEMBER
Set man-trap for milkman.

Purchase quick-dry cement.

Peek at Nigel to see if composting has begun.

Serve scones to police.

SELECTED BIBLIOGRAPHY

Atlas of the Rare Vascular Plants of Ontario. Ed. G.W. Argus, K.M. Pryer, D.J. Whyte, and C.J. Keddy. Ottawa: National Museum of Natural Sciences, 1982-1987.

Avis, Walter S., C. Crate, P. Drysdale, D. Leechman, M.H. Scargill, C.J. Lovell, eds. *A Dictionary of Canadianisms on Historical Principles*. Toronto: Gage, 1967.

Bahlow, Hans. *Deutsches Namenlexicon*. Munich: Suhrkamp Taschenbuchverlag, 1972.

Barney, Stephen A., with Ellen Wertheimer and David Stevens. *Word-Hoard: An Introduction to Old English Vocabulary*. New Haven, CT: Yale University Press, 1977.

Baugh, Albert C., and Thomas Cable. *A History of the English Language*. 3rd ed. London: Routledge and Kegan Paul, 1978.

Ben Abba, Dov. *Signet Hebrew/English, English/Hebrew Dictionary*. New York: New American Library, 1977.

Canadian Encyclopedia. 4 vols. Edmonton: Hurtig, 1988.

Clark, Lewis J. *Wild Flowers of British Columbia*. Sidney, BC: Gray's Publishing, 1973.

Colombo, John Robert, ed. *Colombo's Canadian Quotations*. Edmonton: Hurtig Publishers, 1974.

The Concise Oxford Dictionary of Botany. Ed. Michael Allaby. Oxford: Oxford University Press, 1992.

Cottle, Basil. *The Penguin Dictionary of Surnames*. 2nd ed. Harmondsworth, Eng.: Penguin Books, 1978.

Cowan, David. *An Introduction to Modern Literary Arabic*. Cambridge: Cambridge University Press, 1958.

Downie, Mary Alice and Mary Hamilton, eds. *"And some brought flowers": Plants in a New World*. Toronto: University of Toronto Press, 1980.

Eberhard-Wabnitz, Margit, and Horst Leisering. *Knaurs Vornamenbuch*. Munich: Drömersche Verlagsanstalt, 1985.

Rosa species

Ekwall, Eilert. *The Concise Oxford Dictionary of English Place-Names*. 4th ed. Oxford: Oxford University Press, 1960.

Fucilla, Joseph. *Our Italian Surnames*. New York: Chandler, 1949.

Gage Canadian Dictionary. Rev. ed. Toronto: Gage, 1997.

Genders, Roy. *Bulbs: A Complete Handbook*. London: Robert Hale, 1973.

Gerard, John. *The Herbal or General History of Plants. Revised and enlarged by Thomas Johnson*. London: 1633. Reprinted, New York: Dover, 1989

Grieve, Mrs. M. *A Modern Herbal*. Ed. C.F. Leyel. London: Penguin, 1977.

Harvey, A.G. *Douglas of the Fir*. Cambridge, MA: Harvard University Press, 1947.

Hosie, R.C. *Native Trees of Canada*. 7th ed. Ottawa: Canadian Forestry Service, Information Canada, 1973.

_____. *Native Trees of Canada*. 8th ed. Don Mills, ON: Fitzhenry and Whiteside, 1979.

ITP Nelson Canadian Dictionary of the English Language. Toronto: Nelson, 1997.

Lawrence, George H.M. *Taxonomy of Vascular Plants*. New York: Macmillan, 1951.

Lewis, Charlton T., and Charles Short. *A Latin Dictionary: Founded on Andrew's Edition of Freund's Latin Dictionary*. Impression of 1st ed. 1879. Oxford: Oxford University Press, 1958.

Lewis, Walter H. and Memory P.F. Elvin-Lewis. *Medical Botany: Plants Affecting Man's Health*. New York: John Wiley and Sons, 1977.

Liddell, Henry George, and Robert Scott. *A Greek-English Lexicon*. 9th ed. Oxford: Oxford University Press, 1953.

Merriam-Webster New Book of Word Histories. Ed. F.C. Mish. Springfield, MA: Merriam-Webster, 1991.

Mills, A.D. *A Dictionary of English Place Names*. Oxford: Oxford University Press, 1993.

Morlet, Marie-Thérèse. *Dictionnaire étymologique des noms de famille*. Paris: Perrin, 1991.

Mossé, Fernand. *A Handbook of Middle English*. Trans. James A. Walker. Baltimore, MD: Johns Hopkins Press, 1952.

Mulligan, Gerald A., and Derek B. Munro. *Poisonous Plants of Canada*. Ottawa: Agriculture Canada, 1990.

The National Gardening Association Dictionary of Horticulture. New York: Penguin, 1994.

Naumann, Horst. *Familiennamenbuch*. Leipzig: Bibliographisches Institut, 1989.

A lily from India

The New Shorter Oxford English Dictionary. Oxford: Oxford University Press, 1993.

O'Grady, William and Michael Dobrovolsky. *Contemporary Linguistic Analysis: An Introduction.* 3rd ed. Copp Clark, 1996.

Oxford Companion to the English Language. Ed. Tom McArthur. Oxford: Oxford University Press, 1992.

Oxford English Dictionary. Ed. James A.H. Murray et al. Oxford: Oxford University Press, 1884-1928; corrected reissue, 1933.

Oxford English Dictionary. 2nd ed. Ed. R.W. Burchfield et al. Oxford: Oxford University Press, 1989.

Oxford Russian Dictionary. Rev. ed. Ed. Colin Howlett et al. Oxford: Oxford University Press, 1995.

Partridge, Eric. *A Dictionary of Slang and Unconventional English.* London: Routledge, 1984.

_____. *Origins: A Short Etymological Dictionary of Modern English.* 4th ed. London: Routledge and Kegan Paul, 1966.

Quirk, Randolph, and C.L. Wren. *An Old English Grammar.* 2nd ed. Methuen's Old English Library. London: Methuen, 1957.

Reaney, P.H., and R.M. Wilson. *A Dictionary of English Surnames.* 3rd ed. Oxford: Oxford University Press, 1995.

Schaar, J. van der. *Woordenboek van Voornamen.* Utrecht: Het Spectrum, 1981.

Schama, Simon. *Landscape and Desire.* New York: Knopf, 1995

Scoggan, H.J. *Flora of Canada.* Ottawa: National Museums of Canada, 1978.

Shakespeare, William. *The Oxford Shakespeare.* Ed. W.J. Craig. London: Oxford University Press, 1966.

Shipley, Joseph T. *The Origins of English Words: Discursive Dictionary of Indo-European Roots.* Baltimore, MD: Johns Hopkins University Press, 1984.

Stearn, William T. *Botanical Latin.* New ed. Toronto: Fitzhenry & Whiteside, 1983.

_____. *Stearn's Dictionary of Plant Names for Gardeners: A Handbook on the Origin and Meaning of the Botanical Names of Some Cultivated Plants*. London: Cassell, 1992.

Story, G. M., W.J. Kirwin, J.D.A. Widdowson, eds. *Dictionary of Newfoundland English*. 2nd ed. Toronto: University of Toronto Press, 1990.

Stuart, David, and James Sutherland. *Plants from the Past: Old Flowers for New Gardens*. London: Penguin, 1989.

Taylor, Norman. *Taylor's Encyclopedia of Gardening*. 4th ed. Boston: Houghton Mifflin, 1961.

Theophrastus. *Enquiry into Plants*. 2 vols. Ed. and trans. Sir Arthur Hort. Loeb Classical Library. London: Heinemann, 1916.

Unbegaun, B.O. *Russian Surnames*. Oxford: Oxford University Press, 1972.

Webster's Third New International Dictionary of the English Language. Springfield, MA: G. and C. Merriam, 1976.

White, J. H. *The Forest Trees of Ontario*. Toronto: Ontario Ministry of Natural Resources, 1973.

Whiting, R.E. and P.M. Catling. *Orchids of Ontario: An Illustrated Guide*. Ottawa: Canacoll Foundation, 1986

ILLUSTRATION CREDITS

pp. *i, xx,* 101, Jim Harter, *Men: A Pictorial Archive from Nineteenth-Century Sources,* Dover Publications

pp. *ii, v, vii, xviii, 56, 76, 82, 307,* Carol Belanger Grafton, *1001 Floral Motifs and Ornaments for Artists and Craftspeople,* Dover Publications

pp. *ix, xix,* 135, Carol Belanger Grafton, *Trades and Occupations: A Pictorial Archive from Early Sources,* Dover Publications

pp. *xi,* 19, 52, 99, 115, 118, 121, 123,126, 130, 147, 151, 166, 175, 180, 181, 184, 187, 199, 204, 225, 242, 243, 254, 265, 287, John Gerard, *The Herball or General History of Plants* (1633 Edition)

p. *xii,* Edmund V. Gillon, *Picture Sourcebook for Collage and Decoupage,* Dover Publications

pp. *xxiv,* 3, 37, 39, 42, 45, 66, 69, 71, 72, 73, 78, 85, 90, 160, 317, 318, Emanuel Sweerts, *Early Floral Engravings: from the 1612 Florilegium,* edited by E.F. Bleiler, Dover Publications

p. *xvi,* Carol Belanger Grafton, *Old-Fashioned Christmas Illustrations,* Dover Publications

pp. 1, 192, Carol Belanger Grafton, *Old-Fashioned Romantic Cuts,* Dover Publications

pp.4, 207, 209, 234, 238, 293, 301, 314, 323, 331, 339, Carol Belanger Grafton, *Victorian Floral Illustrations,* Dover Publications

pp. 5, 25, 46, 55, 68, Barry Sherman, *Illustrations of Plants, Shrubs, and Trees,* Dover Publications

pp. 9, 13, 33, 64, 107, 108, 216, 229, 306, 310, Richard G. Hatton, *1001 Plant and Floral Illustrations from Early Herbals,* Dover Publications

pp. 62, 94, 95, 97, 103, 112, 139, 154, 163, 337, Theodore Menten, *Plant and Floral Woodcuts for Designers and Craftsmen,* Dover Publications

p. 256, *Engravings by Hogarth,* edited by Sean Shesgreen, Dover Publications

INDEX

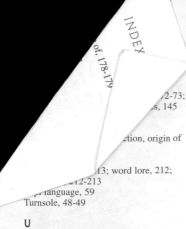